Essays in Biochemistry

volume 34 1999

Essays in Biochemistry

Metalloproteins

Edited by D.P. Ballou

Princeton University Press
Princeton, New Jersey

Published in North America by Princeton University Press,
41 William Street, Princeton, NJ 08540, U.S.A.

Published in the United Kingdom by Portland Press Ltd
on behalf of The Biochemical Society

Portland Press Ltd
59 Portland Place
London W1N 3AJ, U.K.
Fax: 020 7323 1136; e-mail: editorial@portlandpress.com
www.portlandpress.com

Although, at the time of going to press, the information contained in this publication is believed to be correct, neither the authors nor the editor nor the publisher assumes any responsibility for any errors or omissions herein contained. Opinions expressed in this book are those of the authors and are not necessarily held by the Biochemical Society, the editor or the publisher.

ISBN 0-691-05048-1
Library of Congress Catalog Card Number 99-067445

Typeset by Portland Press Ltd
Printed in Great Britain by Information Press Ltd, Eynsham, U.K.

http://pup.princeton.edu

10 9 8 7 6 5 4 3 2 1

Contents

Preface ... xi

Authors ... xv

Abbreviations ... xix

1 The role of efflux in bacterial resistance to soft metals and metalloids
Barry P. Rosen

Introduction ... 1
Soft-metal P-type ATPases ... 3
Resistance to Cu(I) and Ag(I) — CopA and CopB ... 5
Resistance to Zn(II) and Cd(II) — CadA and ZntA ... 7
Resistance to Zn(II) and Cd(II) — CzcCBA ... 8
Resistance to arsenic and antimony ... 9
Conclusions ... 12
Summary ... 13
References ... 13

2 Bacterial detoxification of Hg(II) and organomercurials
Susan M. Miller

Introduction ... 17
Overview of *mer* operons and roles of proteins ... 17
Regulation of gene expression — *merR* ... 18
Proteins of mercury transport ... 22
Organomercurial resistance ... 24
The key step — Hg(II) reduction ... 25
Perspectives ... 28
Summary ... 28
References ... 29

3 Non-haem iron-containing oxygenases involved in the microbial biodegradation of aromatic hydrocarbons

Eric D. Coulter and David P. Ballou

Introduction 31
Oxygen activation at non-haem iron centres 33
Catechol dioxygenases 34
Mononuclear Fe(II) dioxygenases (Rieske oxygenases) 36
Fe(II)–pterin-dependent hydroxylases 41
Di-iron-oxo mono-oxygenases 43
Conclusions 46
Perspectives 46
Summary 47
References 47

4 Haem iron-containing peroxidases

Issa S. Isaac and John H. Dawson

Introduction 51
Compound I 54
Compound II 62
Compound III 63
Compound O 63
Differences in the catalytic activities of P450 and peroxidases 64
Perspectives 65
Summary 66
References 67

5 Nature's universal oxygenases: the cytochromes P450

Stephen G. Sligar

Introduction 71
Classification of P450s 72
Mechanism 76
Perspectives 82
Summary 82
References 82

6 Oxygen-carrying proteins: three solutions to a common problem

Donald M. Kurtz, Jr.

Introduction 85
Reactivity of molecular oxygen with transition-metal ions 86
The active sites of oxygen-carrying proteins 87
Hb and Mb 90
Hr and myoHr 93
Hcy 95
Perspectives 97
Summary 98
References 99

7 Biological electron-transfer reactions

A. Grant Mauk

Introduction 101
Inorganic origins 102
Theoretical basics 103
Electrochemistry and biological electron-transfer kinetics 106
Early biological kinetic experiments 113
Refinements of metalloprotein electron-transfer experiments 116
New challenges concerning intramolecular electron transfer 118
Protein–protein electron-transfer reactions 119
An outlook 121
Summary 121
References 122

8 Molybdenum enzymes

Russ Hille

Introduction 125
Classification of molybdenum enzymes 126
The molybdenum hydroxylases 127
The eukaryotic molybdenum oxotransferases 131
The bacterial molybdenum oxotransferases and related enzymes 133
Perspectives 134
Summary 136
References 136

9 Coenzyme B_{12} (cobalamin)-dependent enzymes

E. Neil G. Marsh

Introduction ... 139
Structures of B_{12} coenzymes ... 140
Interaction of B_{12} with proteins ... 141
AdoCbl-dependent isomerases ... 143
AdoCbl-dependent ribonucleotide reductase ... 147
MeCbl-dependent enzymes ... 148
Perspectives ... 151
Summary ... 152
References ... 153

10 Oxygen reactions of the copper oxidases

James W. Whittaker

Introduction ... 155
O_2 redox chemistry ... 156
Biological chemistry of O_2 ... 159
Copper oxidases ... 160
Role of copper ... 161
Role of the redox cofactor ... 162
Dioxygen reaction with the active site ... 164
Oxygen reactions in cofactor biogenesis ... 168
Summary ... 170
References ... 170

11 Catechol dioxygenases

Joan B. Broderick

Introduction ... 173
Sources of the enzymes and role in bioremediation ... 176
Intradiol catechol dioxygenases ... 177
Extradiol catechol dioxygenases ... 183
Perspectives ... 186
Summary ... 186
References ... 187

12 Cisplatin

Elizabeth E. Trimmer and John M. Essigmann

Introduction....................191
DNA adducts formed by *cis*- and *trans*-DDP....................193
Effects of DNA replication and transcription....................197
Repair of platinum adducts....................198
Recognition of platinum adducts by cellular proteins....................199
Mechanisms of cisplatin resistance....................204
Conclusions and outlook....................208
Summary....................209
Further reading....................209
References....................210

Subject index....................213

Preface

Metals are involved in most segments of the chemistry of life, including respiration, numerous steps of metabolism, photosynthesis, nitrogen fixation, nerve transmission, signal transduction, muscle contraction, oxygen transport and protection from xenobiotic compounds. In addition, metals are used in medicine as therapeutic agents. This volume presents chapters that describe several of these functions of metals in biology.

Although the protein structures of enzymes can provide a wide range of catalytic chemistry, many biological reactions would be inefficient or blocked if the residues of the 20 or so amino acids that are used were the only catalytic tools. For example, reactions involving oxygen are almost non-existent with most amino acids, and electron-transfer reactions would be very inefficient. Nature has thus augmented its catalytic repertoire by incorporating a considerable variety of additional tools. The vitamin cofactors, such as flavins (B_2), niacin (B_1) and cobalamin (B_{12}), carbohydrates and metals all contribute to providing a rich diversity of chemistry required for efficient biological function. Metals allow proteins to react with oxygen, carry out radical reactions and facilitate rapid electron transfer, and they are often excellent in promoting catalysis. In addition, metals such as zinc, calcium and magnesium are crucial to these roles for the maintenance of structures of proteins and nucleic acids. A side benefit of metals is that they serve as spectroscopic handles for helping the researcher monitor various chemical processes. This has led to a multitude of techniques being applicable to the study of metalloproteins in addition to those commonly used for other proteins. These techniques include UV-visible, EPR, electron-nuclear double resonance (ENDOR), NMR, Mössbauer, extended X-ray absorption fine structure (EXAFS), infrared and Raman spectroscopies, magnetic susceptibility and redox measurements, and X-ray crystallography. Most of these techniques will be mentioned in the chapters of this volume.

This volume attempts to show many of the above roles of metals in a variety of circumstances that are used widely in biology. An objective was to present each of the topics from the viewpoint of chemistry. The first two chapters describe processes that bacteria use to rid themselves of toxic metals by either actively pumping them out of the cell or converting them, e.g. mercury, into unharmful forms that can be excreted. Barry Rosen and Susan Miller lucidly present some strategies of how protein side chains can be used to bring about gentle and efficient chemistry that effects detoxification. Chapters 3–6, 8, 10 and 11 are largely devoted to metalloproteins that react with or transport oxygen. Eric Coulter and David Ballou describe a range of non-haem iron-containing oxygenases that are prevalent in the microbial world and which participate in the biodegradation of aromatic compounds. A theme that emerges is that the iron-based active sites in these types of enzymes employ co-ordination environments that exchange ligands during catalysis to permit the various kinds of chemistry to occur. Chapter 4, by Issa Isaac and John Dawson, describes haem iron-containing peroxidases. This chapter covers some of the history of how the mechanisms of peroxidase chemistry were elucidated. A concerted attempt to describe the various oxidation states of the haem prosthetic group provides a solid foundation for understanding the various modes of how iron participates in oxidase and oxygenative chemistry. This chapter serves as an introduction to the 'language' of iron–oxo species, which has been valuable as a framework for thinking about these complex systems. Chapter 5 on cytochromes P450 by Stephen Sligar briefly covers the history of this rich field and captures the spirit of the multiple facets of research on these systems as it builds on the information in Chapter 4. Donald Kurtz describes three very different chemical approaches that have been applied to the transport of oxygen in biological systems. He presents a firm basis of the chemistry and how the structural motifs of proteins are used to bind, but not react irreversibly with, such a strong oxidant as O_2. Grant Mauk (Chapter 6) shows how metalloproteins are so important in carrying out biological electron-transfer reactions. He summarizes theoretical principles in terms that chemists not specializing in physical chemistry can understand, and describes experimental approaches to studying this exciting field. Russ Hille's presentation on molybdenum-containing enzymes (Chapter 8) focuses on oxidative chemistry. He shows how the multiple oxidation states of this oft-neglected metal are well suited for many processes in biology. Neil Marsh writes about how the unique properties of cobalamins promote the formation of metal–carbon bonds whose scission often leads to radical chemistry. This permits a variety of curious carbon-skeleton rearrangements that are vital to metabolism. James Whittaker's chapter on oxygen reactions of the copper oxidases first deals with the unique chemistry of oxygen. His chapter serves well as an introduction to the chemical properties of oxygen. In addition to the mechanisms of catalysis, in several cases, copper enzymes transform themselves by modifying certain residues. This process gives the enzymes a wider range of properties than is provided by the original amino acids. Joan Broderick shows how catechol dioxygenases, which are central to the aerobic biodegradation of most aromatic compounds, catalyse with specificity the cleavage of aromatic rings. The nature of the iron ligation determines what type of cleavage is carried out. Finally, Elizabeth Trimmer and John Essigmann tell us about the chemistry of cisplatin, a metal-containing drug that has been used successfully for the treatment of certain types of cancer. The chemistry of how this drug interacts with

DNA to disrupt the machinery of rapidly dividing cells and the processes whereby cells become resistant to these drugs gives us pause to think about the pharmacological difficulties of dealing with this terrible disease.

This volume is intended mainly for chemistry or biochemistry students in their third or fourth years of undergraduate degrees, or for those beginning postgraduate studies. In addition, many of the articles may be of value to chemists or biochemists who specialize in other fields, and who want to learn what is happening in contemporary bioinorganic chemistry research. The 12 chapters of this volume are written by several of the best-known experts in their specialties. They have tried to present their subjects with a view to teaching principles of chemistry and showing methods of how to develop an understanding of the functions of metals, outlining the roles of metals in a variety of biological functions and presenting perspectives on the future of research in bioinorganic chemistry. No attempt was made to be exhaustive in the coverage of these topics. Only references that appeared crucial to understanding the articles are included, and references to more extensive reviews are given so that readers can go into more depth as desired. The field of bioinorganic chemistry has grown exponentially in the past 20 years, so that only a fraction of this field has been presented here. I would refer the reader to *Principles of Bioinorganic Chemistry* by S.J. Lippard and J.M. Berg (1994), University Science Books, Mill Valley, CA, or to *Chemical Review* (1996), volume **96** (a thematic issue on bioinorganic enzymology), for anyone who wishes to explore these areas further. I hope that the readers of these chapters will begin to feel the excitement of these areas of research and thereby be encouraged to explore them further, and even participate in future research efforts. I am sure that all of the authors will be happy to reply to any questions about their research areas or to receive inquiries about possible research positions.

David P. Ballou
Michigan, 1999

Authors

Barry P. Rosen received his B.Sc. degree from Trinity College, Hartford, CT, U.S.A., in 1965, and his M.Sc. and Ph.D. in Biochemistry from the University of Connecticut in 1968 and 1969 respectively. He was a Public Health Service Postdoctoral Fellow at Cornell University from 1969 to 1971. He became a faculty member in the Department of Biological Chemistry at the University of Maryland School of Medicine in 1972. In 1987 he assumed his current position of Professor and Chairman of the Department of Biochemistry and Molecular Biology at Wayne State University School of Medicine.

Susan M. Miller is currently Assistant Professor of Pharmaceutical Chemistry at the University of California, San Francisco. She received her B.Sc. in Chemistry from the University of Missouri, Columbia, her Ph.D. in Organic Chemistry/Enzymology from the University of California, Berkeley, and postdoctoral training in enzymology at the University of Michigan, Ann Arbor. Her research interests are focused on the study of mechanisms of enzyme catalysis with an interest in both the chemical mechanisms of transformations and how protein-structural features contribute to catalysis. Mechanistic studies of mercuric ion reductase, as well as its interaction with other proteins in the mercury-detoxification pathway, continue to be a significant focus of the laboratory. New areas of interest in the laboratory include the physiological function of a redox-active oestrogen-binding protein from *Candida albicans*, and the mechanism of action of orotate monophosphate decarboxylase, a key enzyme in the uridine biosynthesis pathway.

Eric Coulter is a native of Northern Ireland, where he obtained a B.Sc. in Biological Chemistry from the University of Coleraine in 1991. In 1996, he received a Ph.D. in Chemistry under the direction of Professor John Dawson at the University of South Carolina. Following postdoctoral studies with Professor David Ballou at the University of Michigan, he is currently a postdoctoral associate with Professor Donald Kurtz at the University of Georgia. His current research interests include the mechanism of iron regulation by bacterioferritins and structure-and-function relationships in several mono- and dinuclear non-haem iron proteins.

David P. Ballou received his B.Sc. in Chemistry with a minor in Music from Antioch College in 1965. He did his M.Sc. (1967) and Ph.D. (1971) in Biological Chemistry at the University of Michigan under Graham Palmer. After combined postdoctoral study with M.J. Coon and Vincent Massey, he became an instructor, and is now Professor of Biological Chemistry at Michigan. His research interests are in biological redox systems, especially oxygenations involving flavins and/or metals. His specialities are in physical biochemistry, with an emphasis on spectroscopy and kinetic characterization of intermediates. His other interests include road and mountain biking, downhill and cross-country skiing, music, dancing, sailing and tennis (and most sports).

Issa Isaac was born in Kuwait and his family resides in Palestine. He did his undergraduate studies at Hesston Junior College in Hesston, Kansas, and at Goshen College in Goshen, Indiana, where he obtained a B.A. degree. He then attended Southern Illinois University at Edwardsville and received an M.S. degree in Chemistry. He is a candidate for the Ph.D. degree in Chemistry at the University of South Carolina in Columbia, where he has done his research in the laboratory of Professor John H. Dawson. A significant part of his Ph.D. research has involved a collaborative project that he carried out with Professor David Ballou at the Department of Biological Chemistry at the University of Michigan Medical School in Ann Arbor.

John H. Dawson received an A.B. degree with a major in Chemistry from the University of Columbia and a Ph.D. degree in Chemistry with a minor in Biochemistry from the University of Stanford. He was a National Institutes of Health (NIH) Postdoctoral Fellow in Chemistry at the California Institute of Technology. In 1978, he joined the Department of Chemistry and Biochemistry at the University of South Carolina where he is now Carolina Distinguished Professor with a joint appointment in the School of Medicine. Professor Dawson has been a Camille and Henry Dreyfus Teacher/Scholar, an NIH Research Career Development Awardee and an Alfred P. Sloan Research Fellow. He has been named Outstanding South Carolina Chemist by the SC Section of the American Chemical Society, has received the Russell Award for Research Excellence in Science from the University of South Carolina, and has been elected a Fellow of the American Association for the Advancement of Science. He has recently received the Basic Science Faculty Research Award from the University School of Medicine and has received the Governor's Award for Excellence in Science Discovery from the SC Academy of Science. He was Chair of the Organizing Committee for the Tenth International Conference on Cytochrome P450: Biochemistry, Biophysics and Molecular Biology and is Editor-in-Chief of the *Journal of Inorganic Biochemistry*. Professor Dawson's research interests focus on the structure and function of haem iron oxygenase and peroxidase enzymes and on the application of magnetic circular dichroism to the study of haem proteins. His research is supported by grants from the NIH and NSF.

Stephen G. Sligar received his Ph.D. in Physics from the University of Illinois in 1975. He was an Assistant Professor at Yale University and is now a Professor in the Departments of Chemistry and Biochemistry at the University of Illinois at Urbana-Champaign and a part-time Beckman Institute faculty member in the Advanced Chemical Systems Group. His research is in the area of molecular recognition and the fundamental principles of protein–protein, protein–nucleic-acid and protein–small-molecule interactions using a combination of site-directed mutagenesis, computer modelling and structure determination. His work extends to biomolecular electronics centres, and the synthesis and physical characterization of highly ordered protein superlattices and how they might be useful as building blocks for optically coupled sensors, storage elements and processors. He uses a combination of site-directed mutagenesis, thin-film generation, optical and vibrational polarized spectroscopy, bio-organic chemistry and semi-conductor technology in these endeavours.

Donald M. Kurtz, Jr. received his B.Sc. in Chemistry at the University of Akron in 1972 and his Ph.D. in Physical Biochemistry from Northwestern University under Professor Irving M. Klotz in 1977. After postdoctoral research in the laboratory of Professor Richard H. Holm at Stanford University, he joined the faculty in the Department of Chemistry at Iowa State University in 1979. In 1986 he moved to the University of Georgia where he is currently Professor of Chemistry, Biochemistry and Molecular Biology. His research interests involve the chemistry, enzymology and molecular biology of non-haem iron proteins and enzymes, especially those that interact with oxygen.

Grant Mauk is a Professor of Biochemistry at the University of British Columbia. He obtained his undergraduate degree in Chemistry at Lawrence University in Wisconsin and then completed the M.D.–Ph.D. (Biochemistry) programme at the Medical College of Wisconsin. Following postdoctoral research at the California Institute of Technology, he joined the faculty at the University of British Columbia. His research group studies the kinetic, thermodynamic and spectroscopic properties of genetically modified electron-transfer proteins and the mechanisms by which such proteins form electrostatically stabilized binary complexes.

Russ Hille was born in Tyler, Texas. He did his undergraduate work at Texas Tech University (in Chemistry) and Ph.D. work at Rice University (in Biochemistry, with Dr. John S. Olson). After postdoctoral work at the University of Michigan (in the Department of Biological Chemistry with Dr. Vincent Massey), he joined the faculty of the Department of Medical Biochemistry at the Ohio State University, where he currently holds the position of Professor. His major research interests have been in the application of physical biochemical approaches to studying complex biological redox systems, including molybdenum-containing enzymes and iron–sulphur flavoenzymes.

Neil Marsh received his undergraduate education at Cambridge University and subsequently obtained a Ph.D. in Biochemistry, also from Cambridge. He spent 2 years as a Postdoctoral Fellow in the Chemistry Department at the Johns Hopkins University, Baltimore, before returning as a senior research scientist to the Department of Biochemistry at Cambridge, where he held a Royal Society University Research Fellowship. Since 1995 he has been a member of the faculty in the Department of Chemistry at the University of Michigan.

James Whittaker is an Associate Professor of Biochemistry and Molecular Biology at the Oregon Graduate Institute of Science and Technology in Portland, Oregon. He graduated from the University of Minnesota (Ph.D. in Biochemistry) and was an NIH Postdoctoral Fellow at Stanford University (in the Department of Chemistry). He joined the Chemistry faculty of Carnegie Mellon University in 1986 and moved to Oregon in 1996. His research interests include the electronic structures and dynamics of metalloenzyme active sites and the application of spectroscopic and computational approaches to biomolecular structures.

Joan B. Broderick received a B.Sc. in Chemistry from Washington State University while doing research in both Inorganic Chemistry (with Roger Willett) and Biochemistry (with Tom Okita). She received an N.S.F. predoctoral fellowship to pursue graduate studies in Inorganic Chemistry at Northwestern University (she gained her Ph.D. in 1992). At Northwestern she worked under the direction of Thomas O'Halloran studying spectroscopic and mechanistic aspects of chlorocatechol dioxygenase. She then moved to the Massachusetts Institute of Technology (MIT) as an American Cancer Society Postdoctoral Fellow in the laboratory of Joanne Stubbe, where she investigated the mechanism of the adenosylcobalamin-dependent ribonucleotide reductase. Dr. Broderick spent 5 years on the faculty at Amherst College before moving to Michigan State University in the summer of 1998. Her current research focuses on understanding the role of metal centres, particularly iron–sulphur clusters, in the initiation of radical chemistry.

Elizabeth Trimmer graduated with a B.Sc. degree in Chemistry from Carleton College (Northfield, MN, U.S.A.) in 1988 and obtained her Ph.D. in Biological Chemistry from MIT in 1997. Her doctoral research focused on the interactions of cellular proteins with DNA adducts of the anti-cancer drug cisplatin. She is currently a Postdoctoral Research Fellow in the laboratory of Dr. Rowena Matthews in the Biophysics Research Division and Department of Biological Chemistry at the University of Michigan, Ann Arbor, where she is investigating the mechanism of the flavoenzyme, methylenetetrahydrofolate reductase.

John Essigmann is Professor of Chemistry and Toxicology at MIT. He received his B.Sc. in Biology in 1970 from Northeastern University (Boston, MA, U.S.A.) and S.M. and Ph.D. degrees in Toxicology from MIT in 1972 and 1976, respectively. Following postdoctoral work at MIT, he took a faculty position at the same institution in 1981. His research interests centre on the responses of cells to DNA-damaging agents.

Abbreviations

AdoCbl	adenosylcobalamin
BphC	2,3-dihydroxybiphenyl 1,2-dioxygenase
CCD	chlorocatechol dioxygenase
CCP	cytochrome *c* peroxidase
CCP-ES	original name for compound I of CCP
CPO	chloroperoxidase
CPO-I	CPO compound I
CTD	catechol 1,2-dioxygenase
DDP	diamminedichloroplatinum(II)
ENDOR	electron-nuclear double resonance spectroscopy
ES	enzyme–substrate
EXAFS	extended X-ray absorption fine structure spectroscopy
Hb	haemoglobin
Hcy	haemocyanin
HMG domain	high-mobility group domain
HOMO	highest-occupied molecular orbital
Hr	haemerythrin
HRP	horseradish peroxidase
HRP-I, -II, -III	HRP compounds I, II and III, respectively
LUMO	lowest-unoccupied molecular orbital
Mb	myoglobin
MeCbl	methylcobalamin
Me-H_4-folate	methyltetrahydrofolate
Me-H_4-MPT	methyltetrahydromethanopterin
MFP	membrane fusion protein
MMO	methane mono-oxygenase
myoHr	myohaemrythrin
NDO	naphthalene 1,2-dioxygenase
OMF	outer-membrane factor
PCD	protocatechuate 3,4-dioxygenase
PDO	phthalate dioxygenase
PDR	phthalate dioxygenase reductase
ppt	pyranopterin
RNAP	RNA polymerase
SV40	simian virus 40
TPQ	topaquinone
XP	xeroderma pigmentosum

1

The role of efflux in bacterial resistance to soft metals and metalloids

Barry P. Rosen

Department of Biochemistry and Molecular Biology, Wayne State University School of Medicine, Detroit, MI 48201, U.S.A.

Introduction

All organisms, whether prokaryotic or eukaryotic, have evolved mechanisms that produce resistance to drugs and toxic metals [1,2]. In bacteria, these genes are found both on large multiple-resistance plasmids and on chromosomes. These are most likely to be ancient adaptations to environmental stresses and bacterial warfare. Organisms are exposed constantly to natural sources of metals, including the ionic forms of antimony, arsenic, cadmium, copper, iron, mercury and zinc. In addition, bacteria fight each other by producing antibiotics and toxins that kill competitors, but to which the producing organism is immune. Mankind has adapted this bacterial warfare to its own end by mass production and the use of antibiotics. Which came first, resistance to antibiotics or resistance to metals? Obviously there were dissolved metals in the primordial soup in which the first organisms evolved; still, it is impossible to determine when these resistances first arose. However, metal resistances appear to have been more widespread than antibiotic resistances before the modern era of clinical use of antibiotics.

The most common mechanism of resistance is active extrusion of drugs or toxic metals from the cell, thus reducing their intracellular concentration to subtoxic levels [1]. In this Chapter, various types of extrusion system that produce metal resistance will be considered. Transport can be considered from two points of view; thermodynamics and mechanism. Thermodynamics defines what is possible energetically, but biochemistry defines what molecular events are involved in the transport processes. Transporters that use the same type of energy source are not necessarily similar in terms of biochemical mechanisms. Conversely, transporters that are mechanistically related in terms of evolution may use different energy sources. Thermodynamics divides transporters into primary and secondary systems. *Primary active transport* systems are the biological equivalent of an engine; they utilize chemical energy to establish chemical solute gradients or electrochemical ion gradients. *Secondary active transport* systems are the biological equivalent of a transformer; they convert the electrochemical gradients established by primary systems to gradients of other solutes.

Figure 1 illustrates these categories of transporter with bacterial soft-metal ion transporters. In this context, soft-metal ions are defined as those with high polarizability, where the ratio of the ionic charge to the radius of the ion is relatively large. Thus Na^+ and K^+, with large ionic sizes and single charges, are

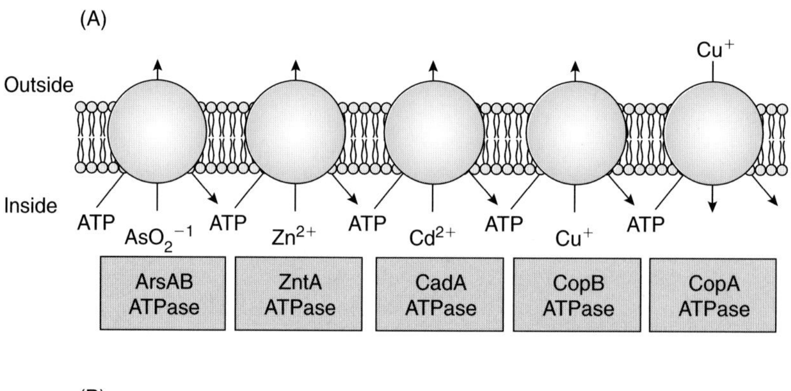

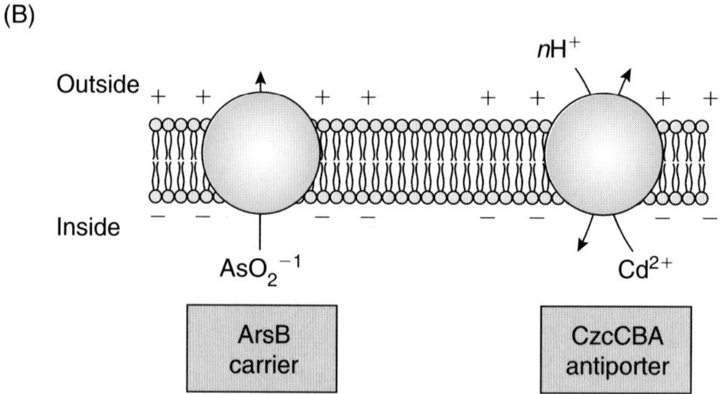

Figure 1. Transport-mediated metal and metalloid resistances
(**A**) Primary pumps catalyse the uptake or extrusion of metal cations or oxyanions. These are ion-translocating ATPases that use the chemical energy of ATP as the driving force for transport. (**B**) Secondary carriers, either uniporters or antiporters, catalyse the electrophoretic uptake or extrusion of cations or oxyanions. The driving forces for transport are electrochemical gradients of protons or other coupling ions established by primary pumps such as respiratory chains.

hard-metal ions. In contrast, ions such as Cu(I), Ag(I), Cd(II) and Hg(II) are soft-metal ions. For the purposes of this discussion, Zn(II) and the metalloids, As(III) and Sb(III), can also be considered soft metals. Although sometimes termed heavy metal transporters, many of these systems utilize substrates such as zinc, which is not a heavy metal. Ions with similar electronic configurations have similar chemical properties. Thus metal ions within the same group are chemically similar. For example, each of the pairs; Cu(I)/Ag(I), As(III)/Sb(III) or Zn(II)/Cd(II), are transported by a single pump or carrier. A number of different and unrelated families of ion pumps and carriers have evolved, each the product of separate evolution.

Many genes related to metal-transport systems have been identified, but only in a few cases have their protein products been demonstrated to catalyse energy-dependent efflux reactions. These include three related cation-translocating ATPases: the *copB* system of *Enterococcus hirae* that encodes a Cu(I)-translocating ATPase for resistance to Cu(I) and Ag(I) [3–5]; the plasmid-encoded *cad* operon of *Staphylococcus aureus* that encodes resistance to Cd(II) [6], and the *zntA* gene that confers zinc and cadmium resistance in *Escherichia coli* by encoding a Zn(II)-translocating ATPase [7]. These ATP-coupled transporters are primary pumps. In contrast, the *czc* determinant from *Alcaligenes eutrophus* [8,9], which also produces resistance to Zn(II) and Cd(II), as well as Co(II), functions as a secondary cation–proton antiporter that is dependent on the electrochemical gradient across the cell membrane. Antiporters bring about concerted movement of two species in opposite directions (as described later). Finally, a resistance system encoded by the *ars* operon of the *E. coli* plasmid R773 produces resistance to arsenate [As(V)], arsenite [As(III)] and antimonite [Sb(III)] [2,10,11]. The Ars transport system is novel in that it can utilize either ATP or the electrochemical gradient [12]; depending on its subunit composition, it can be either a primary pump or a secondary carrier [13,14].

Soft-metal P-type ATPases

One superfamily of cation-translocating ATPases is the E_1E_2 or P-type ATPase group [15] (Figure 2). Members of this family have been found in all organisms, and nearly all biologically important cations are transported by these types of pumps. In animals, the best known P-type ATPases are the sodium pump, which exchanges intracellular Na^+ for extracellular K^+, and is the major electrogenic pump of mammalian cells, the H^+/K^+-ATPase, which exchanges intracellular H^+ for extracellular K^+, and is responsible for gastric acidification, and the calcium-extrusion pumps of plasma membrane and sarcoplasmic reticulum, which are involved in a variety of functions including muscle contraction and intracellular signalling. In bacteria, some P-type ATPases catalyse uptake of essential cations such as Mg^{2+} and K^+. All P-type ATPases have certain common features, such as an ATP-binding domain and a conserved aspartate residue that becomes phosphorylated during the catalytic cycle.

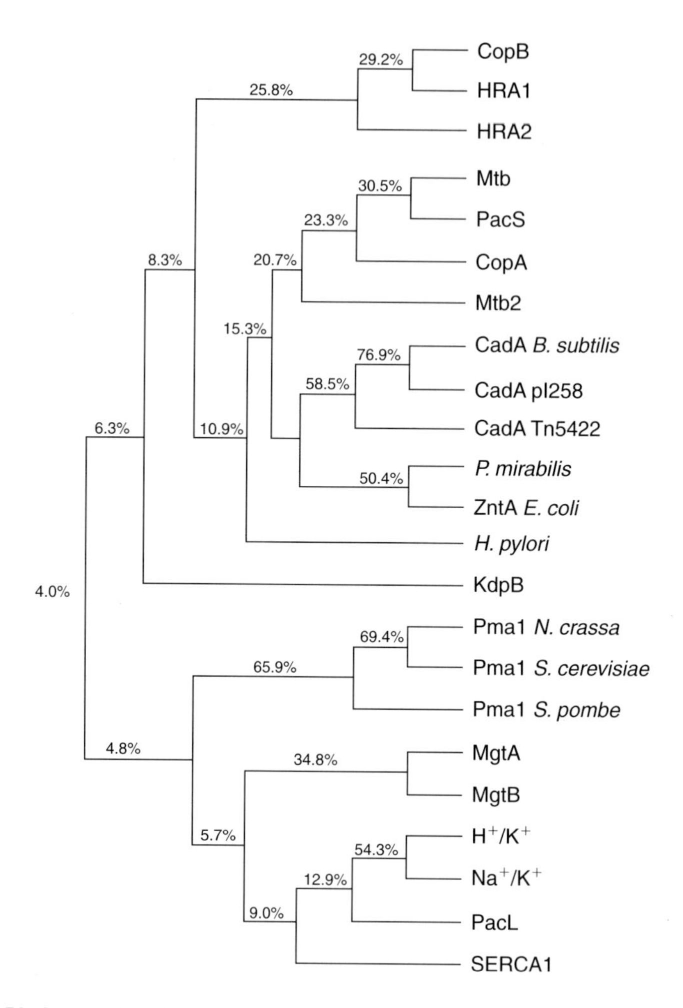

Figure 2. Phylogenetic relationships of the amino acid sequences of soft-metal-translocating P-type ATPases

The dendogram was made using DNasis (Hitachi). The calculated matching percentages are indicated at each branch point. Proteins (known functions are identified; otherwise reading frames are identified solely from DNA sequence analysis): CopB, *En. hirae* copper pump; HRA1 and HRA2, reading frames originally identified as *E. coli* genes; Mtb and Mtb2, *Mycobacterium tuberculosis* reading frames; PacS, *Synechococcus* reading frame; *H. pylori*, *H. pylori* reading frame; *P. mirabilis*, *P. mirabilis* reading frame; ZntA, *E. coli* zinc pump; CadA pI258, *Staph. aureus* plasmid pI258 cadmium pump; CadA Tn5422, *Staph. aureus* transposon reading frame; CadA *B. subtilis*, *B. subtilis* transposon reading frame; CopA, *En. hirae* copper-uptake pump; KdpB, *E. coli* potassium-uptake pump; Pma1, fungal plasma membrane proton pumps (of *Neurospora crassa*, *Saccharomyces cerevisiae* and *Schizosaccharomyces pombe*); MgtA and MgtB, *Salmonella typhimurium* magnesium-uptake pumps; H^+/K^+, mammalian gastric proton pump; Na^+/K^+, mammalian sodium pump; PacL, *Synechococcus* reading frame; SERCA1, mammalian sarcoplasmic reticulum calcium pump.

A subfamily of the P-type ATPases is that which has been termed the CPx-type ATPases [5] or soft-metal ATPases [7]. These are involved in intracellular homoeostasis or resistance to soft-metal ions, which are almost invariably toxic in high concentrations. Some, such as Cu(I) and Zn(II), are required for viability because they are cofactors for enzymes or regulatory proteins. Thus there are transport systems that accumulate these ions inside cells. However, if allowed to accumulate to high levels, they would become toxic. Specific soft-metal P-type ATPases have evolved to keep the intracellular concentrations of these ions within a tolerable range by the process of homoeostasis. As an additional benefit to the cells, copper pumps can detoxify the more deadly soft-metal Ag(I), and the zinc pump detoxifies Cd(II).

Resistance to Cu(I) and Ag(I) — CopA and CopB

Cu(I) reacts strongly with sulphur- and nitrogen-donor ligands in proteins. Thus even at low concentrations, it is a very effective cofactor in enzymes. However, it can also react non-specifically with cysteine and histidine residues in proteins, disrupting their enzymic activity. Transition-metal ions such as Cu(I) can also produce reactive oxygen species, which can lead to various modes of cellular damage. Thus copper is extremely toxic at high concentrations. One branch of the evolutionary tree of soft-metal P-type ATPases includes Cu(I)-translocating enzymes (Figure 2). These have been found in nearly every organism examined, reflecting the biological necessity for copper homoeostasis. In humans, mutations in the genes for two of these enzymes result in inherited defects in copper metabolism, termed Menkes and Wilson diseases [16–18]. Examination of the primary and predicted secondary structures of copper ATPases demonstrates several common features absent in other P-type ATPases. First, they all have a cysteine- or histidine-rich N-terminus. In the eukaryotic homologues, such as the Menkes and Wilson disease proteins, there are six repeats of a cysteine-X-X-cysteine sequence, where X can be any of a variety of amino acid residues. In bacterial homologues such as CopA from *En. hirae*, there is only one CXXC sequence [3]. In CopB, another bacterial homologue from *En. hirae*, there are no CXXC motifs, but the N-terminus is rich in histidine residues [19]. The N-termini of these proteins is predicted to be a cytosolic extension that binds copper ions. The bound copper could then be transferred to a domain in the membrane for transport. Alternatively, the N-terminal domains could be regulatory, with binding of metal activating transport of a different ion of copper. Another possibility is that the N-terminal domain sequesters copper to prevent it from entering the cytosol as a free ion. At present there are no data to discriminate between these possibilities.

The substrate of copper P-type ATPases is probably Cu(I), although it is difficult to demonstrate this conclusively. CopB has been shown to transport ^{64}Cu(I) *in vitro* [4]. Under oxidizing conditions, such as exposure to the atmosphere, Cu(I) can rapidly oxidize to Cu(II). However, the cytosol of cells is highly reducing, with high concentrations of thiols such as reduced glutathione, suggesting that Cu(I) is present intracellularly. Recently, Lutsenko and co-workers [20] cloned the N-terminal sequences from the Menkes and

Wilson proteins and demonstrated that they bound copper with a stoichiometry of 5–6 nmol of copper/nmol of protein (one copper per CXXC repeat) when expressed in *E. coli*. The copper released from polypeptides reacted with bicinchoninic acid, which reacts with Cu(I) but not Cu(II). These results support but do not prove the hypothesis that the reduced metal ion is the pump substrate.

In the middle of each of these proteins is a conserved cysteine-proline-cysteine (CPC) or cysteine-proline-histidine (CPH) motif. Transport proteins are polytopic membrane proteins; that is, they span a membrane multiple times. From the hydropathic profile of these proteins it can be predicted that they will have eight or more membrane-spanning regions. The topology of a putative copper ATPase from *Helicobacter pylori* was determined recently [21]. The ATPase has eight domains that probably form four pairs of transmembrane segments. The ATP-binding domain is located in a cytoplasmic sequence between transmembrane helices 6 and 7. Transmembrane helix 6 contains both the conserved aspartate that is the most likely phosphorylation site and the CPC sequence characteristic of the soft-metal-translocating ATPases. The function of the CPC sequence is not known, but it is reasonable to speculate that it is part of the translocation pathway for the ion.

The physiological function of bacterial copper transporters has been best elucidated by the studies of Solioz and Vulpe [5]. They cloned and sequenced the *copAB* operon from *En. hirae* [3,19]. The *copA* gene product is a 727-residue protein with one CXXC motif in the N-terminus and a CPC sequence in a putative membrane-spanning helix. It is more closely related to the Menkes protein than it is to CopB. CopB lacks N-terminal CXXC repeats but is rich in histidine; it also has a transmembrane CPH sequence. Since histidine nitrogens can serve as Cu(I) ligands, those residues may serve the same role in CopB as the cysteine repeats in CopA. Disruption of the *copA* gene resulted in a copper-requiring phenotype. In contrast, disruption of *copB* rendered the cells sensitive to copper. This has led to the hypothesis that CopA is a copper-uptake system and CopB is an efflux system [4]. Consistent with this idea, CopB has been shown to catalyse ATP-dependent accumulation of ^{64}Cu(I) and ^{110}Ag(I) in everted (inside-out) membrane vesicles prepared from cells of *En. hirae* [4]. Transport was inhibited by vanadate, a classical inhibitor of P-type ATPases. It is reasonable to consider that together the two pumps maintain the intracellular concentration of copper within a narrow range, sufficient for synthesis of copper-containing proteins but at subtoxic levels. The similarities and differences between CopA and CopB illustrate an important point: the direction of substrate transport is not an intrinsic feature of transport proteins and is not easily deduced from inspection of the sequence — similar proteins can function either as uptake or efflux pumps.

Resistance to Zn(II) and Cd(II) — CadA and ZntA

Zn(II) is not as polarizable as Cu(I), and so does not interact with thiols or nitrogens in proteins as strongly as copper ions. As a consequence, zinc ions are not as toxic as copper ions. Still, both are required in low amounts and toxic in high amounts. In contrast, cadmium, a heavy metal in the same group as zinc, is extremely toxic, and cadmium-resistance determinants are wide-spread. The first Cd(II) resistance was identified in the *Staph. aureus* plasmid pI258. The *cad* operon contains two overlapping open reading frames [6]. The first, CadC, a member of the ArsR family of metalloregulatory proteins, is a cadmium/zinc-responsive repressor protein that most likely controls expression of the *cad* operon. The *cadA* gene, which overlaps *cadC* by 4 bp, encodes a 727-residue P-type ATPase. From this sequence similarity, CadA was proposed to confer resistance by transporting cadmium out of cells. CadA has phosphorylation (DKTGT) and phosphatase (TGES) domains common to all members of the family. The first 105 residues are probably cytosolic and have a single CXXC sequence. In a putative transmembrane helix is the sequence $C_{371}PC_{373}$. The *cad* operon was subcloned from plasmid pI258 and expressed in *Bacillus subtilis* [22]. The operon could be induced by cadmium, and its expression conferred cadmium resistance to the *B. subtilis* host cells. Cadmium transport catalysed by CadA was demonstrated using everted membrane vesicles of *B. subtilis* [22]. Transport was assayed by measuring the uptake of ^{109}Cd(II) into the vesicles. Cd(II) transport was observed only when the plasmid contained a functional *cadA* gene, and when ATP was used as an energy source.

In contrast to Gram-positive organisms such as *Staph. aureus* and *B. subtilis*, the Gram-negative *E. coli* is intrinsically resistant to cadmium. Although it was possible that *E. coli* once had a chromosomal *cadA* gene, when the sequencing of the *E. coli* genome was completed, it was apparent that *cadA* was not present in the chromosome. A homologue, *orf732*, was more closely related to *cadA* than to *copA* or *copB*. However, ORF732 and CadA share only 35% identical residues, so it was not possible to reliably assign a function to the putative protein. In a search for zinc-responsive genes in *E. coli*, Beard et al. [23] isolated a zinc- and cadmium-sensitive mutant by transposon mutagenesis, calling the disrupted gene *zntA*. From its sequence, *zntA* was shown to be identical to *orf732*. In a parallel study, Rensing et al. [7] disrupted *orf732* and showed that the resulting strain was sensitive to both zinc and cadmium. Since the gene product is clearly related to zinc metabolism, the designation ZntA has been accepted. Everted membrane vesicles from a wild-type strain accumulated ^{65}Zn(II) and ^{109}Cd(II) using ATP as an energy source. Transport was sensitive to the classical P-type ATPase inhibitor vanadate. Membrane vesicles from the *zntA*-disrupted strain accumulated neither ^{65}Zn(II) nor ^{109}Cd(II). Both the sensitive phenotype and transport defect of the mutant were complemented by expression of *zntA* on a plasmid. This was the first demonstration of zinc transport by a soft-metal P-type ATPase [7]. Recently, a ZntA homologue was identified in *Proteus mirabilis*, an enteric bacterium related to *E. coli*. A mutant in this ZntA homologue was identified from a defect in swarming of *P. mirabilis* [24]. The relationship of a putative zinc-extrusion pump and swarming is not clear. Perhaps an increase in intracellular zinc ions results in inhibition of an enzyme involved in swarming.

Although the chromosomal zinc pumps are closely related to the cadium pumps, their normal functions are probably different. Since zinc is required for growth but is toxic in excess, a balance between uptake and efflux is necessary. ZntA is probably the efflux half of a zinc homoeostatic mechanism, similar to the situation with CopA and CopB. In contrast, plasmid-encoded pumps such as CadA most likely evolved from such housekeeping genes, but have a specialized role in resistance mechanisms.

Resistance to Zn(II) and Cd(II) — CzcCBA

A system evolutionarily distinct from the P-type ATPase that also produces resistance to Cd(II) and Zn(II) is encoded by the plasmid-borne *czc* (cadmium-zinc-cobalt) operon of the Gram-negative bacterium *A. eutrophus* [25]. Czc is a member of a family of exporters that provide resistance to drugs and metals, among other functions [26]. These exporters are a complex of three types of protein: (i) a cytoplasmic membrane-export system; (ii) a membrane-fusion protein (MFP); and (iii) an outer-membrane factor (OMF). The three structural genes of the operon encode the 116 kDa CzcA (integral membrane-export protein), 55 kDa CzcB (MFP) and 45 kDa CzcC (OMF) proteins. This membrane complex is a secondary transport system that catalyses extrusion of Cd(II), Zn(II) or Co(II) in exchange for an undefined number of protons [8]. These secondary transport systems are active only in the sense that uphill transport of one solute occurs, but at the expense of a gradient of another, with a net decrease in the sum of the two gradients. There are three basic types of secondary porter: uniporters, symporters and antiporters. As the names imply, these are carriers for single solutes, two or more solutes in the same direction, or multiple solutes in opposite directions, respectively. The Czc system is therefore an antiporter; the driving force for metal extrusion is the downhill flow of protons, where the electrochemical proton gradient had been established previously by primary proton pumps.

Partial resistance is conferred by expression of the *czcA* gene alone, suggesting that CzcA is sufficient for metal extrusion [9]. From its amino acid sequence deduced from the DNA sequence, CzcA is a member of the resistance-nodulation-cell division (RND) family [26]. From its hydropathic profile, CzcA should be an integral membrane protein and may be the central component of the transporter. The primary sequence of CzcB places it in the MFP family. From its hydropathic profile CzcB probably has at least one membrane-spanning region near the N-terminus that may embed it in the inner membrane. From analysis of gene fusions, it appears that CzcB is accessible to the periplasm. When inner and outer membranes were separated, the majority of CzcB was found associated with the outer membrane. This suggests that its function may be to create a bridge for cations from the inner-membrane CzcA antiporter across the periplasm and through the outer membrane. CzcB has two similar histidine-rich motifs near its N-terminus (HGDTEHH and HGDGEHH). Since histidine nitrogens are ligands to soft metals, it was proposed that these regions would be involved in metal binding or recognition. However, deletion of both motifs led only to a small loss of resistance, so the role of these histidines in Czc function remains unclear. From its primary sequence, CzcC is a member of the OMF family. The results of gene-fusion experiments indicate that CzcC is a periplasmic protein that is also associated with the outer membrane. Deletion of the *czcC* gene reduces but does not eliminate zinc resistance, although there is a significant reduction in the level of cadmium resistance, so it is possible that CzcC is a specificity factor. Another possible role of CzcC is interaction of the complex with an outer-membrane pore-forming protein, facilitating the efflux of cations, possibly forming the pore itself; thus CzcC would not be essential for resistance, but could augment it [9].

Resistance to arsenic and antimony

A third type of metal resistance is conferred by bacterial *ars* operons (Figure 3) [2]. They share a common organization, with genes for a transcriptional repressor, ArsR, that regulates expression of the operon, a transport protein, ArsB, and a soluble reductase, ArsC. Arsenate [As(V)] is reduced by the ArsC arsenate reductase to arsenite [As(III)], which is then extruded by the transport system. This expands the range of resistance to both the oxidized and reduced forms of arsenic.

ArsB is the actual carrier protein for arsenite or antimonite. It functions as a secondary uniporter (a carrier for a single solute) that catalyses electrophoretic transport of the oxyanion out of cells [14]. The ArsB protein from the *E. coli* plasmid R773 has 12 membrane-spanning α-helices, with a transmembrane topology similar to that of many secondary carriers [27]. It has no required cysteine residues and apparently transports arsenite or antimonite as anions rather than as soft metals [28]. The R773 ArsB is closely related to most other bacterial arsenic-resistance transporters (Figure 4). However, recently a separate group of proteins with ArsB-like function has been identified in eukaryotic yeast, archaea and some bacteria. From their hydropathic profiles these proteins may have only 10 membrane-spanning segments. It is not clear whether these are related to the more common ArsB or whether they represent a parallel evolutionary solution to the same problem of arsenic toxicity.

Operons that have only *arsRBC* genes produce low levels of resistance to As(III) or Sb(III). Some *ars* operons confer higher-level resistance. These

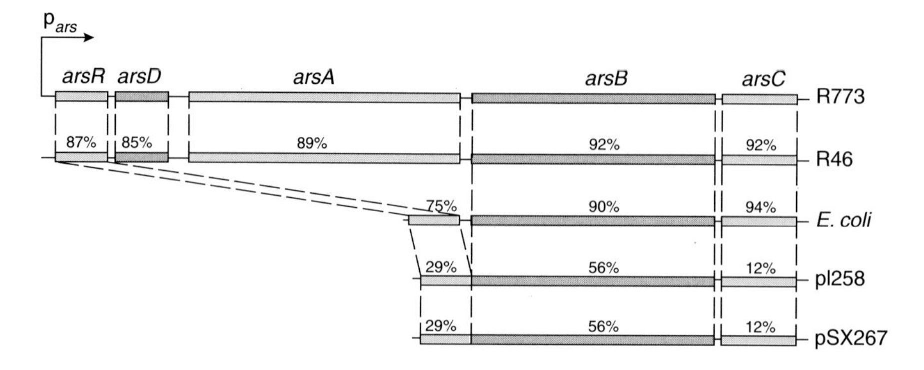

Figure 3. Bacterial *ars* operons

The five genes of the *ars* operon of *E. coli* plasmid R773 are shown with the direction of transcription indicated by the arrow, starting with the promoter, p_{ars}. Genes are indicated by boxes, with the intergenic spaces as single lines. The genes of homologous *ars* operons of *E. coli* plasmid R46, the *E. coli* chromosomal operon, and the Staphylococcal plasmids pI258 and pSX267 are aligned below. The similarities of the gene products to the R773 proteins are given as percentage identity.

operons have two additional genes; *arsD*, which encodes a second regulatory protein [29], and *arsA*, which encodes an ATPase protein that associates with ArsB to convert it into a primary ATP-coupled arsenite pump [12,13]. The two proteins form a membrane-bound complex that cannot be easily dissociated. However, in the absence of an *arsB* gene, ArsA can be expressed as a soluble protein in the cytosol of *E. coli*, which can be easily purified [30]. ArsA ATPase activity is low in the absence of arsenite or antimonite, ensuring that ATP hydrolysis does not occur in the absence of transport. ATPase activity is allosterically activated by As(III) or Sb(III). These bind as soft metals to a site on ArsA that contains three cysteine residues, Cys113, Cys172 and Cys422, with the postulated structure shown in Figure 5(A) [31,32]. This is the first physiologically essential metal-binding site involving arsenic or antimony, and ArsA is the first enzyme shown to require arsenic or antimony for activity. ArsA has two halves, each of which has a binding site for ATP. Cys113 is located in the analogous position in the N-terminal half of the protein as Cys422 occupies in the C-terminal half. The working hypothesis for allosteric activation is that the two halves of the protein must be in contact with each other to be active (Figure 5B) [33]. In the absence of the soft metal there is nothing to hold them together, so the activity is low. When As(III) or Sb(III) interact with cysteines from the two halves of the protein, those domains are pulled together, resulting in activation of ATPase activity. The release of energy from ATP hydrolysis in the ArsA subunit is transduced into the ArsB subunit of the pump. ArsB is postulated to bind arsenite or antimonite anion — different molecules than those bound at the allosteric site of ArsA — and to use the energy from hydrolysis to pump those toxic metalloid oxyanions out of the cells.

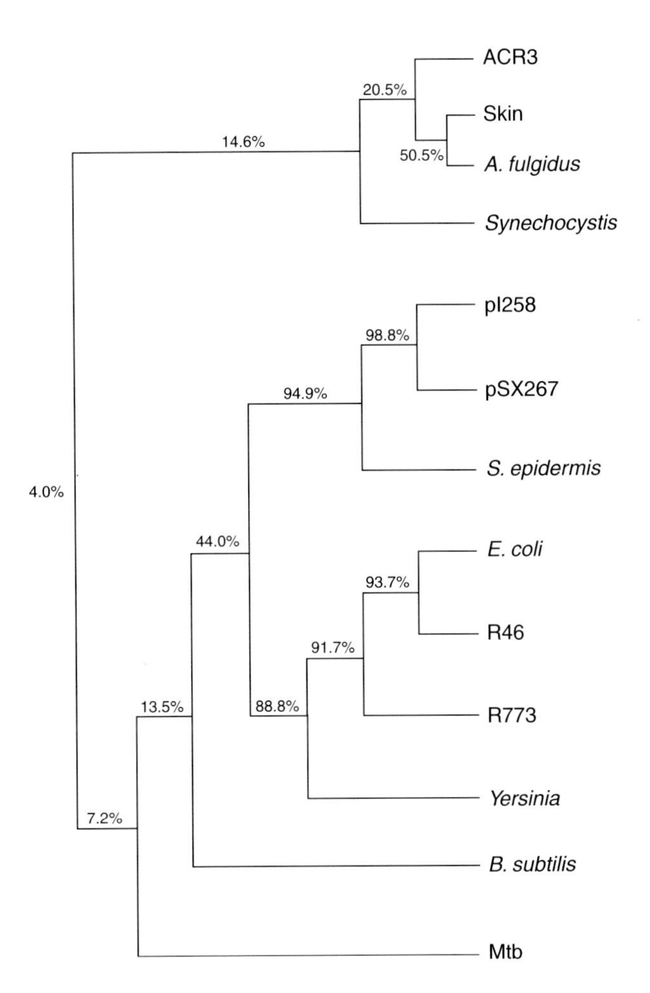

Figure 4. Phylogenetic relationships of the amino acid sequences of arsenic carrier proteins of prokaryotes, archaea and eukaryotes
Proteins: ACR3, *S. cerevisiae* (yeast) arsenite-resistance protein; Skin, *B. subtilis* SKIN element (transposon) arsenite-resistance protein; *A. fulgidus*, *Archaeoglobus fulgidus* (archaea) reading frame; *Synechocystis*, *Synechocystis* sp. PCC6803 reading frame; pI258, *Staph. aureus* plasmid pI258 arsenite carrier; pSX267, *Staph. aureus* plasmid pSX267 arsenite carrier; *S. epidermis*, *Staph. epidermis* reading frame; *E. coli*, *E. coli* chromosomal arsenite carrier; R46, *E. coli* plasmid R46 arsenite carrier; R773, *E. coli* plasmid R773 arsenite carrier; *Yersinia*, *Yersinia pestis* reading frame; *B. subtilis*, *B. subtilis* reading frame; Mtb, *M. tuberculosis* reading frame.

Why should there be two different modes of arsenic transport? It is reasonable to consider that ArsB arose early, giving a moderate level of resistance to arsenic salts. More operons with *arsB* alone have been identified to date than operons with both *arsA* and *arsB*. However, pumps are thermodynamically more efficient than carriers. With a few assumptions as to the stoichiometries of the systems, ArsB alone is theoretically capable of producing a 1000-fold gradient of arsenite; that is, the concentration of arsenite inside of a resistant cell would be 1000-fold less than the medium concentration. In

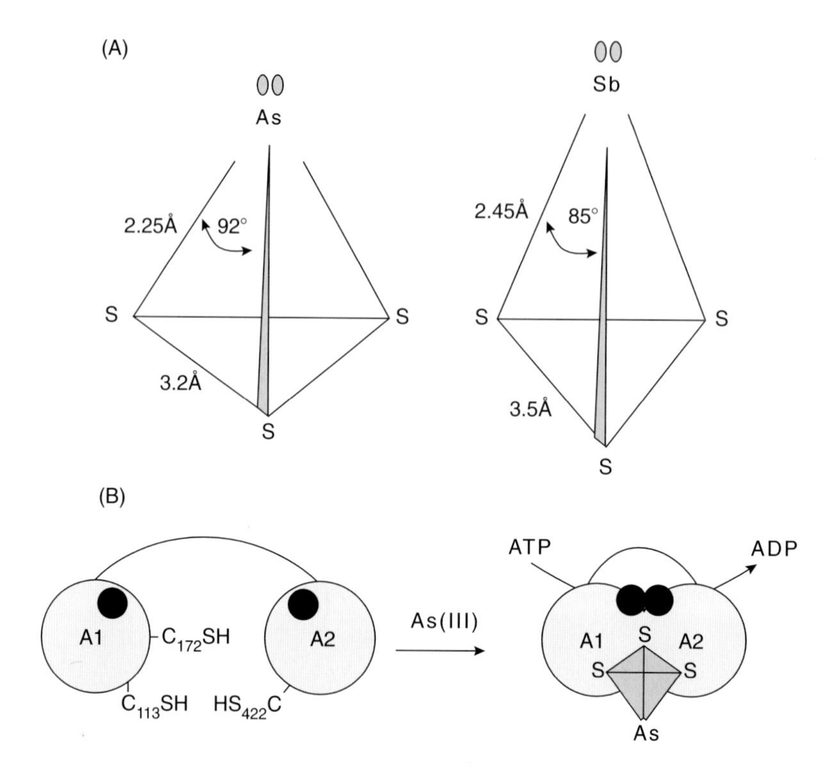

Figure 5. Arsenic-binding sites in proteins
(**A**) Geometry of the allosteric As(III)- and Sb(III)-binding site in the ArsA ATPase. In ArsA the allosteric site contains Cys113, Cys172 and Cys422. The trigonal pyramidal structure contains three co-ordinately ligated sulphur thiolates, with As(III) or Sb(III) at the apex. The bond angles and distances are postulated from crystallographic analysis of small molecules containing As–S or Sb–S bonds. (**B**) Model for interaction of nucleotide-binding sites in the ArsA ATPase. The 583-amino acid residue ArsA protein has two analogous halves, the N-terminal A1 and the C-terminal A2 halves. Both A1 and A2 halves have ATP binding sites (●). Neither site is catalytically active in the absence of the allosteric activators As(III) or Sb(III). When metalloid is bound at the allosteric site, the two domains of the protein are brought together such that an A1 domain from one subunit forms an interface with an A2 domain from the other subunit. This brings the two ATP-binding sites into close proximity, promoting catalysis.

contrast, an ArsA–ArsB pump could, in theory, produce a 10^6-fold gradient! This can go far to explain why cells that express both genes are more resistant than those with only *arsB*. Thus the evolution of the ArsA–ArsB complex was a later event that may have arisen in response to chronic exposure to elevated levels of arsenic. It is interesting to speculate that other pumps may have evolved from secondary carriers [34].

Conclusions

Resistance to metals produced by efflux systems is probably the most frequently adopted mechanism used in Nature. As a result of the genome-sequencing project, many potential systems have been identified from DNA sequences. In the future the functions of these systems will be elucidated. In some cases they will be primarily resistance systems such as those found on plasmids or chromosomes as the result of gene transfer. In other cases those pumps will have housekeeping functions in normal metabolism, such as copper or zinc metabolism. Humans also have such housekeeping systems, and inheritable metabolic disorders of copper homoeostasis such as Menkes and Wilson diseases result from mutations in the genes for such pumps. It is advantageous for genetic and biochemical studies to use bacteria, and they serve as excellent model systems for the study of human diseases. For example, in soft-metal P-type ATPases, the roles of the N-terminal cysteines and histidines and the membrane CPC domain are not known, and the structure of the metal-binding sites are yet to be elucidated. Using bacterial pumps, the combination of molecular genetics and biochemistry can shed light on the way in which metals are bound and transported by their human counterparts.

Summary

- *Bacteria have evolved various types of resistance mechanism to toxic soft metals and metalloids, including cadmium/zinc, copper/silver and arsenic/antimony.*
- *Active efflux of the metal is a frequently utilized stratagem to produce resistance by lowering the intracellular concentration to subtoxic levels.*
- *Reduction to a less-toxic form or to a form recognized by an efflux system also occurs.*
- *Pumps utilized for resistance may have evolved from normal cellular systems. For example, plasmid-mediated cadmium resistances may have evolved from a common ancestor of the pump involved in zinc homoeostasis.*
- *Pumps are more efficient than carriers and may have evolved by developing carriers that associate with ATPase subunits.*

References

1. Dey, S. & Rosen, B.P. (1995) Mechanisms of drug transport in prokaryotes and eukaryotes. In *Drug Transport in Antimicrobial and Anticancer Chemotherapy* (N.H. Georgopapadakou, ed.), pp. 103–132, Dekker, New York
2. Rosen, B.P. (1996) Bacterial resistance to heavy metals. *J. Biol. Inorg. Chem.* **1**, 273–277
3. Odermatt, A., Suter, H., Krapf, R. & Solioz, M. (1992) An ATPase operon involved in copper resistance by *Enterococcus hirae*. *Ann. N.Y. Acad. Sci.* **671**, 484–486
4. Solioz, M. & Odermatt, A. (1995) Copper and silver transport by CopB-ATPase in membrane vesicles of *Enterococcus hirae*. *J. Biol. Chem.* **270**, 9217–9221
5. Solioz, M. & Vulpe, C. (1996) CPx-type ATPases: a class of P-type ATPases that pump heavy metals. *Trends Biochem. Sci.* **21**, 237–241
6. Nucifora, G., Chu, L., Misra, T.K. & Silver, S. (1989) Cadmium resistance from *Staphylococcus aureus* plasmid pI258 *cadA* gene results from a cadmium-efflux ATPase. *Proc. Natl. Acad. Sci. U.S.A.* **86**, 3544–3548

7. Rensing, C., Mitra, B. & Rosen, B.P. (1997) The *zntA* gene of *Escherichia coli* encodes a Zn(II)-translocating P-type ATPase. *Proc. Natl. Acad. Sci. U.S.A.* **94**, 14326–14331
8. Nies, D.H. (1995) The cobalt, zinc, and cadmium efflux system CzcABC from *Alcaligenes eutrophus* functions as a cation-proton antiporter in *Escherichia coli*. *J. Bacteriol.* **177**, 2707–2712
9. Rensing, C., Pribyl, T. & Nies, D.H. (1997) New functions for the three subunits of the CzcCBA cation-proton antiporter. *J. Bacteriol.* **179**, 6871–6879
10. Chen, C.M., Misra, T.K., Silver, S. & Rosen, B.P. (1986) Nucleotide sequence of the structural genes for an anion pump. The plasmid-encoded arsenical resistance operon. *J. Biol. Chem.* **261**, 15030–15038
11. Tsai, K.J., Hsu, C.M. & Rosen, B.P. (1997) Efflux mechanisms of resistance to cadmium, arsenic and antimony in prokaryotes and eukaryotes. *J. Inst. Zool. Acad. Sinica. Taipei* **36**, 1–16
12. Dey, S. & Rosen, B.P. (1995) Dual mode of energy coupling by the oxyanion-translocating ArsB protein. *J. Bacteriol.* **177**, 385–389
13. Dey, S., Dou, D. & Rosen, B.P. (1994) ATP-dependent arsenite transport in everted membrane vesicles of *Escherichia coli*. *J. Biol. Chem.* **269**, 25442–25446
14. Kuroda, M., Dey, S., Sanders, O.I. & Rosen, B.P. (1997) Alternate energy coupling of ArsB, the membrane subunit of the Ars anion-translocating ATPase. *J. Biol. Chem.* **272**, 326–331
15. Fagan, M.J. & Saier, Jr., M.H. (1994) P-type ATPases of eukaryotes and bacteria: sequence analyses and construction of phylogenetic trees. *J. Mol. Evol.* **38**, 57–99
16. Bull, P.C. & Cox, D.W. (1994) Wilson disease and Menkes disease: new handles on heavy-metal transport. *Trends Genet.* **10**, 246–252
17. Vulpe, C., Levinson, B., Whitney, S., Packman, S. & Gitschier, J. (1993) Isolation of a candidate gene for Menkes disease and evidence that it encodes a copper-transporting ATPase. *Nat. Genet.* **3**, 7–13
18. Solioz, M., Odermatt, A. & Krapf, R. (1994) Copper pumping ATPases: common concepts in bacteria and man. *FEBS Lett.* **346**, 44–47
19. Odermatt, A., Suter, H., Krapf, R. & Solioz, M. (1993) Primary structure of two P-type ATPases involved in copper homeostasis in *Enterococcus hirae*. *J. Biol. Chem.* **268**, 12775–12779
20. Lutsenko, S., Petrukhin, K., Cooper, M.J., Gilliam, C.T. & Kaplan, J.H. (1997) N-terminal domains of human copper-transporting adenosine triphosphatases (the Wilson's and Menkes disease proteins) bind copper selectively *in vivo* and *in vitro* with stoichiometry of one copper per metal-binding repeat. *J. Biol. Chem.* **272**, 18939–18944
21. Melchers, K., Weitzenegger, T., Buhmann, A., Steinhilber, W., Sachs, G. & Schafer, K.P. (1996) Cloning and membrane topology of a P-type ATPase from *Helicobacter pylori*. *J. Biol. Chem.* **271**, 446–457
22. Tsai, K.J., Yoon, K.P. & Lynn, A.R. (1992) ATP-dependent cadmium transport by the cadA cadmium resistance determinant in everted membrane vesicles of *Bacillus subtilis*. *J. Bacteriol.* **174**, 116–121
23. Beard, S.J., Hashim, R., Membrillo-Hernandez, J., Hughes, M.N. & Poole, R.K. (1997) Zinc(II) tolerance in *Escherichia coli* K-12: evidence that the *zntA* gene (*o732*) encodes a cation transport ATPase. *Mol. Microbiol.* **25**, 883–891
24. Lai, H.C., Gygi, D., Frasier, G.M. & Hughes, C. (1998) A swarming-defect mutant of *Proteus mirabilis* lacking a putative cation-transporting P-type ATPase. *Microbiology* **144**, 1957–1961
25. Nies, D.H., Nies, A., Chu, L. & Silver, S. (1989) Expression and nucleotide sequence of a plasmid-determined divalent cation efflux system from *Alcaligenes eutrophus*. *Proc. Natl. Acad. Sci. U.S.A.* **86**, 7351–7355
26. Paulsen, I.T., Park, J.H., Choi, P.S. & Saier, Jr., M.H. (1997) A family of Gram-negative bacterial outer membrane factors that function in the export of proteins, carbohydrates, drugs and heavy metals from Gram-negative bacteria. *FEMS Microbiol. Lett.* **156**, 1–8
27. Wu, J., Tisa, L.S. & Rosen, B.P. (1992) Membrane topology of the ArsB protein, the membrane subunit of an anion-translocating ATPase. *J. Biol. Chem.* **267**, 12570–12576

28. Chen, Y., Dey, S. & Rosen, B.P. (1996) Soft metal thiol chemistry is not involved in the transport of arsenite by the Ars pump. *J. Bacteriol.* **178**, 911–913
29. Chen, Y. & Rosen, B.P. (1997) Metalloregulatory properties of the ArsD repressor. *J. Biol. Chem.* **272**, 14257–14262
30. Rosen, B.P., Weigel, U., Karkaria, C. & Gangola, P. (1988) Molecular characterization of an anion pump. The *arsA* gene product is an arsenite(antimonate)-stimulated ATPase. *J. Biol. Chem.* **263**, 3067–3070
31. Bhattacharjee, H., Li, J., Ksenzenko, M.Y. & Rosen, B.P. (1995) Role of cysteinyl residues in metalloactivation of the oxyanion-translocating ArsA ATPase. *J. Biol. Chem.* **270**, 11245–11250
32. Bhattacharjee, H. & Rosen, B.P. (1996) Spatial proximity of Cys113, Cys172, and Cys422 in the metalloactivation domain of the ArsA ATPase. *J. Biol. Chem.* **271**, 24465–24470
33. Li, J., Liu, S. & Rosen, B.P. (1996) Interaction of ATP binding sites in the ArsA ATPase, the catalytic subunit of the Ars pump. *J. Biol. Chem.* **271**, 25247–25252
34. Rosen, B.P., Dey, S., Dou, D., Ji, G., Kaur, P., Ksenzenko, M., Silver, S. & Wu, J. (1992) Evolution of an ion-translocating ATPase. *Ann. N.Y. Acad. Sci.* **671**, 257–272

2

Bacterial detoxification of Hg(II) and organomercurials

Susan M. Miller

Department of Pharmaceutical Chemistry, University of California, San Francisco, CA 94143–0446, U.S.A.

Introduction

Environmental toxins can be grouped loosely as organic or inorganic (i.e. metal-ion-containing) compounds. Man-made organic toxins are often persistent in Nature because their unnatural chemical structures are resistant to existing biochemical mechanisms of detoxification. By contrast, toxic metal-ion-containing compounds are often naturally occurring, but have been re-dispersed and concentrated in the environment through human activities. Because of the natural exposure to metal ions, many micro-organisms long ago elaborated mechanisms to avoid their toxicity. The key to these mechanisms is prevention of metal-ion complexation of important functional groups on biomolecules (e.g. proteins and DNA), a goal that has been accomplished in many ways, including sequestration by specific chelators, exportation by specific transporters, and redox conversion to less toxic oxidation states [1]. In this Chapter I will examine Nature's design of the most common detoxification pathway identified for mercuric ion compounds.

Overview of *mer* operons and roles of proteins

Although some organisms have developed strategies to sequester Hg(II), most employ a redox strategy for detoxification that takes advantage of the physiologically accessible reduction potential for Hg(II) to Hg(0). Being hydrophobic and volatile, Hg(0) can escape from the bacterial cell, leaving it unharmed. The genes encoding this detoxification pathway are frequently organized in regulated operons, but are not always found on plasmids associated with transposons carrying antibiotic-resistance genes. Figure 1(A) summarizes several operons from both Gram-negative and Gram-positive bacteria [2,3]. Although the organization varies, common genes in all operons include: *merR*, which codes for an Hg(II) sensing, DNA-binding, regulatory protein; *merA*, which codes for the mercuric-ion-reducing protein; and one or more of the genes *merP*, *merT* and/or *merC*, which code for proteins involved in transporting mercuric ions into the cytoplasm of the bacterial cell. Organisms possessing this minimum set of genes exhibit a narrow-spectrum resistance, meaning they can survive in the presence of mercuric-ion compounds with exchangeable ligands. Several operons also contain a gene designated *merB*, which codes for an organomercurial lyase protein that catalyses the cleavage of covalent C–Hg bonds. Organisms possessing these operons exhibit a broad-spectrum resistance to both organomercurials and mercuric-ion compounds. The additional gene designated *merD* appears to code for a protein that binds DNA weakly, and may serve a co-regulatory, but non-essential, role. Those genes designated as open reading frames have yet to be functionally characterized.

Assignment of the basic functions for the pathway proteins comes in large part from genetic analyses of mutations and deletions in the Tn*21* operon [4]. Subsequent cloning and expression of individual genes, primarily from Tn*501*, Tn*21* and *Bacillus* sp. RC607, have allowed detailed biochemical analyses of the individual functions of several of the proteins. Figure 1(B) summarizes the overall cellular organization of the pathway proteins in Gram-negative bacteria. The same organization occurs in Gram-positive bacteria; however, with no outer membrane or periplasmic space, it is unclear whether a secreted MerP is involved. In the sections that follow, our current understanding of the structures and functions of the individual proteins is described. A recurring theme to note is the presence of essential cysteine residues in each of the proteins. This is to be expected, since thiol-containing compounds exhibit the highest affinity for Hg(II) and, in fact, have long been referred to by the generic name mercaptans, meaning mercury capturing.

Regulation of gene expression — *merR*

As alluded to above, synthesis of the detoxification proteins only occurs when cells are exposed to Hg(II) (or organomercurials in the case of broad-spectrum resistance). As with most inducible pathways, control of protein expression occurs at the level of gene transcription (mRNA synthesis), and involves binding of the protein coded by the *merR* gene to the operator/promoter region of the *mer* operon. Extensive studies of the interactions of the MerR proteins from Tn*21* and Tn*501* with their respective DNA operator/promoter sequences have revealed several unusual features of the interaction [5]. As indicated in Figure 2(A), the *merR* gene in these operons is transcribed divergently

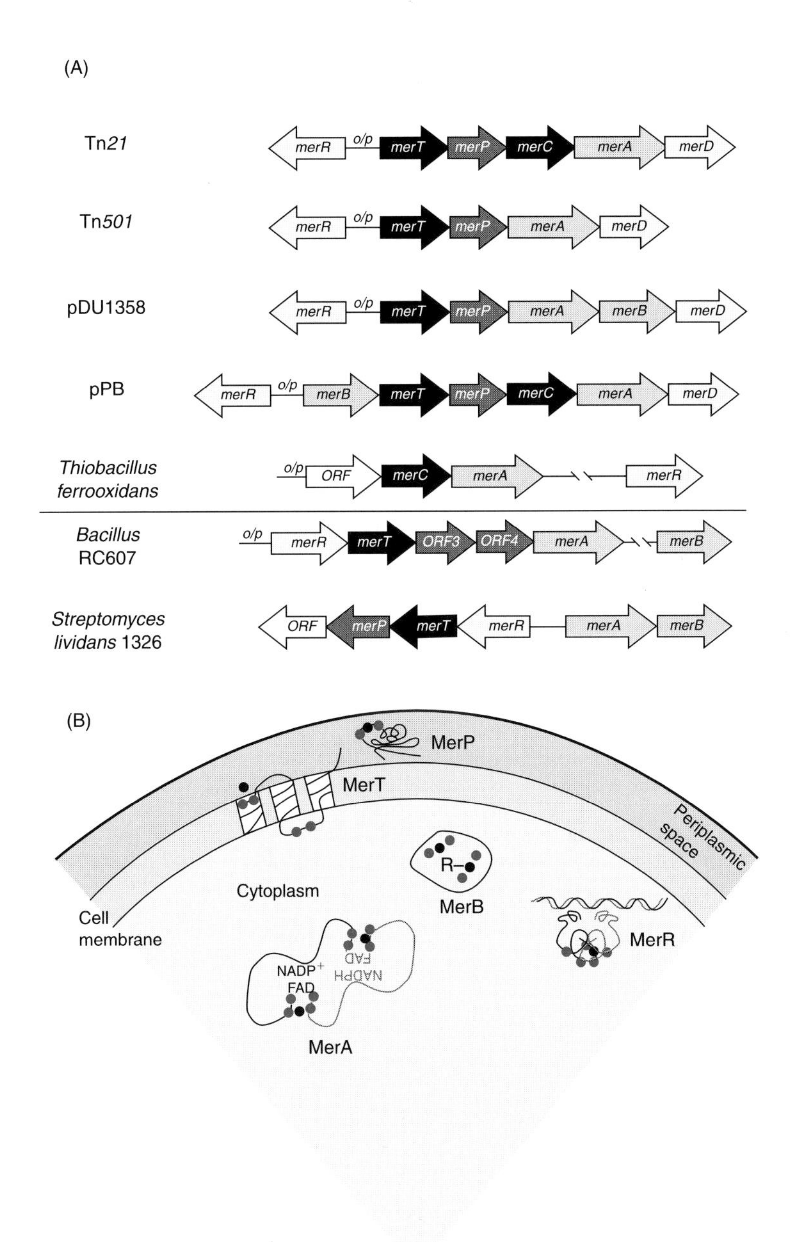

Figure 1. Organization of *mer* operons and Mer proteins
(**A**) Organization of several *mer* operons from Gram-negative (above the line) and Gram-positive (below) bacteria [2]. Bacterial sources for the top four include: *Shigella flexneri* and *Escherichia coli* (Tn*21*), *Pseudomonas aeruginosa* (Tn*501*), *Serratia marcescens* (pDU1358), and *P. stutzeri* (pPB). Sequences from *Thiobacillus ferrooxidans*, *Bacillus* sp. RC607 and *Streptomyces lividans* are chromosomally encoded. o/p, operator/promoter region. Adapted in part from [2], with permission from Elsevier Science, and in part from [3], with permission from BirkHäuser Verlag AG. (**B**) Cellular organization of Mer proteins in Gram-negative bacteria. Blue dots indicate essential cysteine residues in each protein, black dots indicate Hg(II) ions, and R–● is an organomercurial.

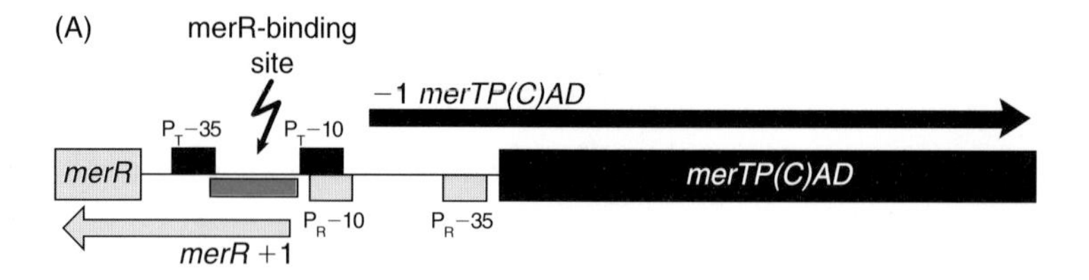

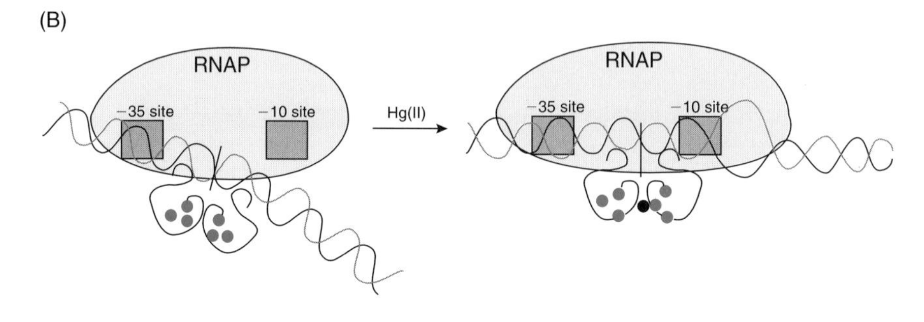

Figure 2. Interaction of MerR with the o/p region of the Tn*21* and Tn*501* mer operons

(**A**) Generalized view of the divergently transcribed Tn*21* and Tn*501* operons [4]. Horizontal arrows indicate the directions of transcription from each promoter. Adapted from [5] with permission. ©1992 American Society for Microbiology. (**B**) Repression [-Hg(II)] and activation [+Hg(II)] of transcription from the P_T promoter (−35 and −10 sites) in the MerR–DNA–RNAP complex [5]. Adapted with permission from *Nature (London)* [6]. © (1995) Macmillan Magazines Ltd.

from the detoxification genes [*TP(C)AD*]; hence, there are two divergent, but overlapping, promoter sequences, P_R and P_T, for binding of RNA polymerase (RNAP) and subsequent transcription in opposite directions. Neither promoter has the optimal 17-bp spacing between the RNAP recognition sequences, but transcription from P_R is preferred in the absence of the MerR protein. This makes sense because MerR is the Hg(II) sensor and the other proteins are not needed unless the toxin is sensed. MerR binds as a dimer to a symmetrical stretch of DNA between the -35 and -10 RNAP-recognition sequences in the P_T promoter. Since this site overlaps the transcriptional start site for the *merR* gene itself, the MerR protein represses its own transcription as well as that of the detoxification genes.

One unusual feature of the MerR protein–DNA complex is the finding that RNAP still binds to the -35 site of the P_T promoter when MerR is present. Although atypical of most repressors, this behaviour is ideal in a detoxification system, because it effectively primes the system for immediate transcription of the detoxification genes at the slightest influx of the highly toxic Hg(II).

A second unusual feature of the protein is that upon binding of the inducer ligand Hg(II), MerR neither dissociates completely from nor moves to a new site on the DNA. Instead, the Hg(II)–MerR complex activates transcription of the detoxification genes while remaining bound at the same location and while still repressing transcription of its own gene. Extensive evidence indicates that it does this by altering the shape of the DNA. In the absence of Hg(II), the MerR dimer induces a double kink in the DNA at its recognition sequences [6]. Binding of Hg(II) alters the protein–DNA interaction just enough to release the bend in the DNA and allow an untwisting or underwinding of the DNA in the suboptimal spacer in the P_T promoter. This relaxation of the kink in the DNA, coupled with the underwinding of the spacer, is envisioned to allow the pre-associated RNAP to bind to its -10 recognition site, allowing transcription of the detoxification genes. Figure 2(B) is a cartoon of this current model for repression and activation [6].

The remaining feature of the MerR protein to consider is the nature of the Hg(II)-binding site. As the sensor for this highly toxic species, the Hg(II)-binding site is expected to exhibit both high selectivity and high affinity in order to compete with the many binding sites for Hg(II) in the cell. MerR is indeed exquisitely sensitive, activating transcription *in vitro* with only nanomolar concentrations of Hg(II) in the presence of 10^6-fold higher concentrations of competing thiol-containing ligands, exactly as needed to compete in the thiol-rich cytoplasm of the bacterial cell. Likewise, the protein shows a 10^2–10^3-fold selectivity for Hg(II) over similar metal ions, like Cd(II), Zn(II) and Ag(I) [7].

How does the protein achieve this high affinity and selectivity? Clearly, cysteine thiols are the expected ligands, but the question is: how many and in what geometry? In solution, linear Hg(II)–dithiol complexes with relatively short Hg–S bonds are most common. Although of high affinity, these also rapidly exchange with other thiols by association/dissociation of a third ligand. With sterically hindered thiol ligands, 3- and 4-co-ordinate Hg(II)–thiol complexes with slightly longer Hg–S bonds are favoured and have higher formation constants overall. This suggests that a higher-affinity Hg(II)–protein complex might utilize more than two cysteines positioned on the semi-rigid scaffold of the protein to optimize the Hg–S bond lengths and geometry for a 3- or 4-co-ordinate complex. In MerR, only one Hg(II) binds per dimer, again the most efficient design, since the dimer is the active DNA-binding unit. Comparison of several MerR protein sequences indicates that three cysteines are completely conserved in the monomers, yielding six cysteines available for the single Hg(II) per dimer. By comparison with the spectral properties of several crystallographically characterized 2-, 3- and 4-co-ordinate Hg–thiol model complexes, O'Halloran and co-workers showed that Hg(II)–MerR is a complex with *three* cysteine residues [8,9]. To examine which three participate, Helmann and co-workers generated a series of single, and multiple, Cys→Ala mutations of the three conserved cysteines in the *Bacillus* RC607 protein [10]. Their studies indicate that Hg(II) forms an asymmetric complex across the dimer interface using one conserved cysteine from one monomer and the other two conserved cysteines from the other monomer. Presumably, formation of this asymmetric Hg(II) complex requires a significant conformational change that drives the observed structural changes in the protein–DNA complex. Elucidation of the disposition of the three cysteines and the nature of the conformational change awaits information on the three-dimensional structures of MerR [± Hg(II)] and its DNA complex.

Proteins of mercury transport

As illustrated in Figure 1(B), the actual detoxification step, i.e. reduction of Hg(II) to Hg(0), occurs within the cytoplasm of the bacterial cell and, thus, requires that Hg(II) enter the cell. Non-specific entry of mercuric ions into cells is neither rapid nor high in flux. Thus to avoid the toxic effects of Hg(II) binding to the exterior of cells, each operon codes for at least one protein involved in transporting Hg(II) specifically across the inner cell membrane.

Among the characterized Gram-negative operons, the *merT*/*merP* combination is most common and clearly sufficient for resistance. Of this pair, *merT* codes for an integral membrane protein with two pairs of cysteines, which provides the actual cross-membrane transport, whereas *merP* codes for a soluble protein with one pair of cysteines and is secreted into the periplasmic space. The *merC* gene codes for another integral membrane protein that appears to be the sole transport protein in *Thiobacillus ferrooxidans*. However, deletion of *merC* in the Tn*21* system shows little effect on resistance [11]. A role of enhancing resistance has been suggested based on the higher occurrence of *merC* genes in bacteria isolated from mercury-polluted sediments [2]. In the Gram-positive operons, the only obvious transport gene is homologous to *merT*.

The presence of both a periplasmic protein and a transporter in Gram-negative species is reminiscent of nutrient importers, where the functions of the two proteins are highly coupled. In those systems, little or no transport occurs unless the periplasmic protein is present to scavenge the low levels of nutrients and deliver them to the transporter through specific protein–protein interactions. Although a similar role may be envisioned for MerP, the toxicity of Hg(II) suggests that a protective role may be equally or more important. Studies of the Tn*501* system suggest that the functions of the two proteins are not highly coupled [12]. Thus when the *merT* gene is expressed in the absence of any other *mer* operon genes, the cells become hypersensitive to Hg(II), meaning they die at lower Hg(II) concentrations than cells expressing no *mer* operon genes. This indicates that transport of Hg(II) into the cell occurs readily in the absence of MerP. When both *merP* and *merT* are expressed, the cells are slightly less hypersensitive, consistent with an increased binding capacity for Hg(II) due to the presence of the MerP protein. This suggests a major role for MerP as a scavenger to protect other periplasmic and membrane proteins from binding Hg(II), rather than as an enhancer of transport.

Additional studies using site-directed Cys→Ser mutations of MerP and MerT have provided further insight into their functions and interactions. In

MerP, both cysteines are required for specific binding of Hg(II) [13]. In the absence of the first cysteine in the sequence, the second one remains reduced [12,13], and the expressed protein inhibits transport in the Tn501 system [12]. This suggests that MerP does specifically interact with MerT to hand off bound Hg(II), even though it is not essential for transport to occur. Structures of MerP have recently been reported {oxidized [14]; and reduced ± Hg(II) [15], and see Figure 3}, providing the opportunity to design further mutagenesis studies to probe the specificity determinants for this MerP–MerT interaction. In MerT, there are two pairs of conserved cysteines: the first pair appears in the first of three predicted transmembrane helices, and the second pair is predicted to appear in a cytoplasmic loop. Since nothing is known of the oligomeric state of MerT in the membrane, or whether the two pairs of cysteines in a single monomer could approach one another in three dimensions, the nature of the Hg(II)-binding site(s) is unclear. However, mutation of either

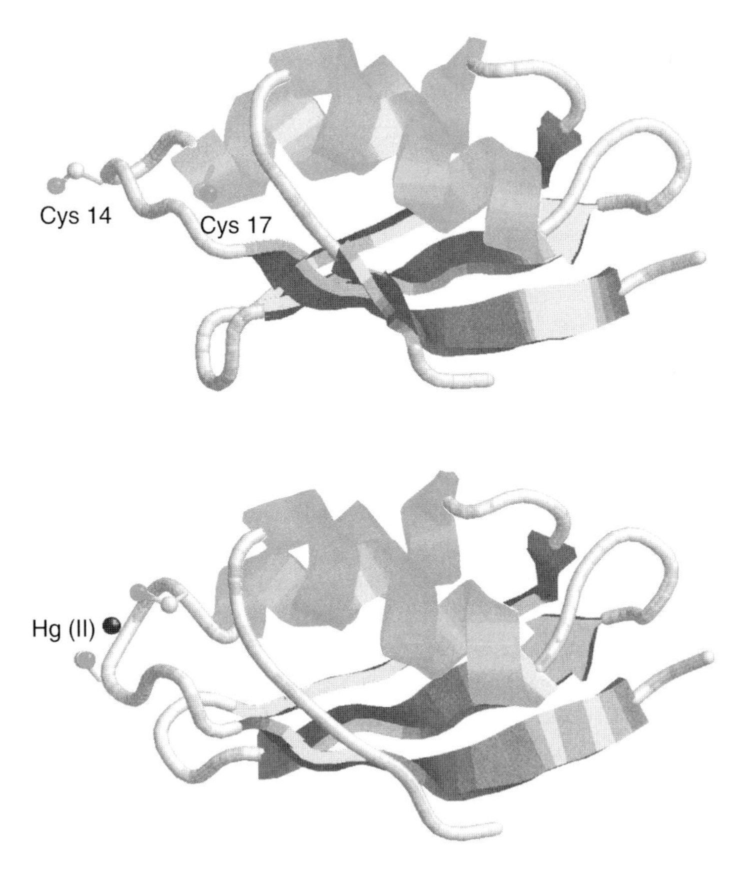

Figure 3. Structure of reduced MerP (Tn2*1*) in the absence (top) and presence (bottom) of Hg(II) [15]
Blue dots are the cysteine sulphurs and the black dot is Hg(II). Structures were generated using RasMac v2.6 (freeware) and co-ordinates from Brookhaven Protein Databank files 1afi (model 2) and 1afj (model 7) for the top and bottom structures, respectively.

cysteine in the transmembrane pair to serine leads to complete loss of Hg(II) transport; whereas the second pair appears to be less critical for transport [12]. Further analysis of the role of these residues is needed.

Organomercurial resistance

As described above, broad-spectrum resistance to organomercurials, in addition to mercuric ions, is associated with the presence of only one additional gene, *merB*, which codes for an organomercurial lyase. Since MerR represses expression of all the detoxification genes, it must also be responsive to organomercurials. Although not extensively studied, it appears that removal of 15–20 of the C-terminal amino acids of MerR from operons containing a *merB* gene reduces or eliminates resistance to organomercurials, suggesting this portion of MerR may be important for sensing organomercurial compounds [5]. Transport of organomercurials is also not well characterized, but it appears that small hydrophobic organomercurials like methylmercury diffuse non-specifically across the membrane, whereas others, such as phenylmercury, require transport by MerT [16] for efficient flux.

The reaction catalysed by organomercurial lyase involves cleavage of the C–Hg bond to yield a hydrocarbon and a mercuric ion with exchangeable ligands (eqn. 1) that can then serve as a substrate for mercuric ion reductase:

$$R_3C-Hg-SR+RSH \longrightarrow R_3C-H+Hg(SR)_2 \qquad (1)$$

In vitro assays of the lyase require thiol compounds as ligands for activity. The protein is quite promiscuous, accepting aromatic or aliphatic, large or small substrates. This has allowed a fairly thorough study of the stereochemistry and mechanism of C–Hg bond cleavage using a variety of substrates, including several probes for generation of radical intermediates [17]. The results indicate complete retention of stereochemistry and integrity of chemical structure, indicating no formation of radical intermediates in the reaction. Overall, the results are most consistent with an S_E2 (bimolecular electrophilic substitution) mechanism (Figure 4) involving a proton transfer to the incipient carbanion, concomitant with nucleophilic attack at Hg(II) by an enzyme residue. Analysis of several *merB* sequences indicates there are three highly conserved cysteines in the protein. Site-directed mutagenesis of any of these cysteines to alanine in MerB in the *Escherichia coli* R831 operon led to a loss of resistance to phenylmercuric acetate, suggesting that all of the cysteines play a role in liganding the R_3C–HgX substrate and the Hg(II) product before it is released [18]. Although not characterized, it seems likely that the Hg(II) product of the lyase would be transferred directly to mercuric ion reductase *in vivo* for the final detoxification step.

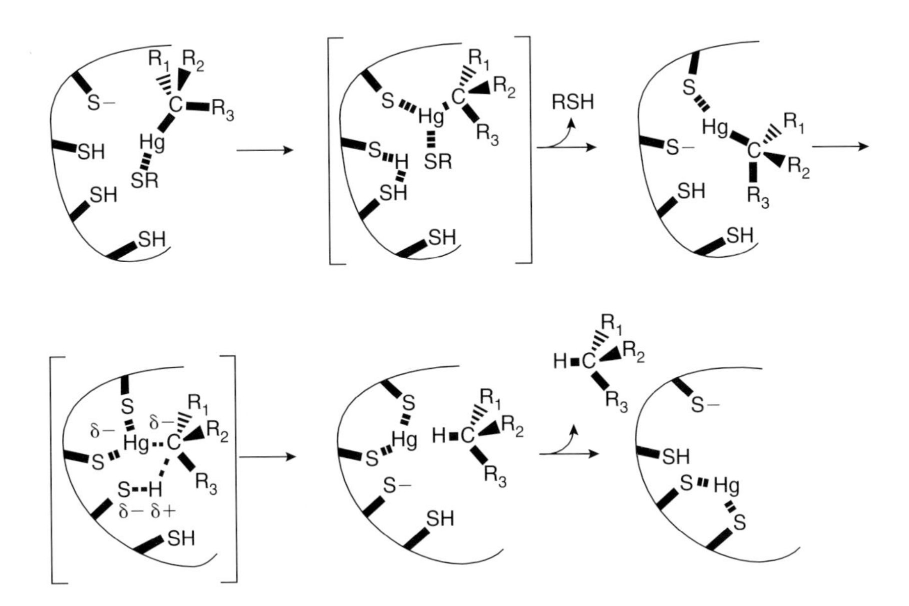

Figure 4. Hypothetical mechanism for organomercurial lyase
Roles are proposed for all four essential enzyme cysteine residues.

The key step — Hg(II) reduction

Upon arrival in the cytoplasm, mercuric ions undergo the actual detoxification step, two-electron reduction to elemental mercury. As shown in eqn. 2, NADPH is the source of electrons [19], and because of the high thiol content of the cell, an Hg–dithiol complex is the expected form of the substrate:

$$NADPH + Hg(SR)_2 + H^+ \longrightarrow NADP^+ + Hg(0) + 2RSH \qquad (2)$$

Mercuric ion reductase (MerA), which catalyses the reaction, is a homodimeric enzyme with two interfacial active sites [20]; see Figure 5(A). Structurally, it is a homologue of the flavin-disulphide oxidoreductase enzymes [21,22], such as glutathione reductase or lipoamide dehydrogenase, which catalyse reversible redox reactions involving NAD(P)H and a disulphide substrate. The common feature of their active sites is the presence of an electron-mediating FAD cofactor sandwiched between the binding site for NAD(P)H and a redox-active pair of cysteines. In the disulphide reductases, these cysteines cycle between oxidized and reduced states, but in MerA they remain reduced and form part of the Hg(II)-binding site [23]. Unique to MerA is a second pair of cysteines [24], contributed by the C-terminus of the other monomer, and two tyrosines, one from each monomer, which complete the residues in each Hg(II)-binding site (Figure 5B).

The presence of multiple cysteines in the Hg(II)-binding sites of MerA is consistent with its need to trap Hg(II); but very tight complexation, as observed in MerR, is not desirable here, since tight complexation would lower

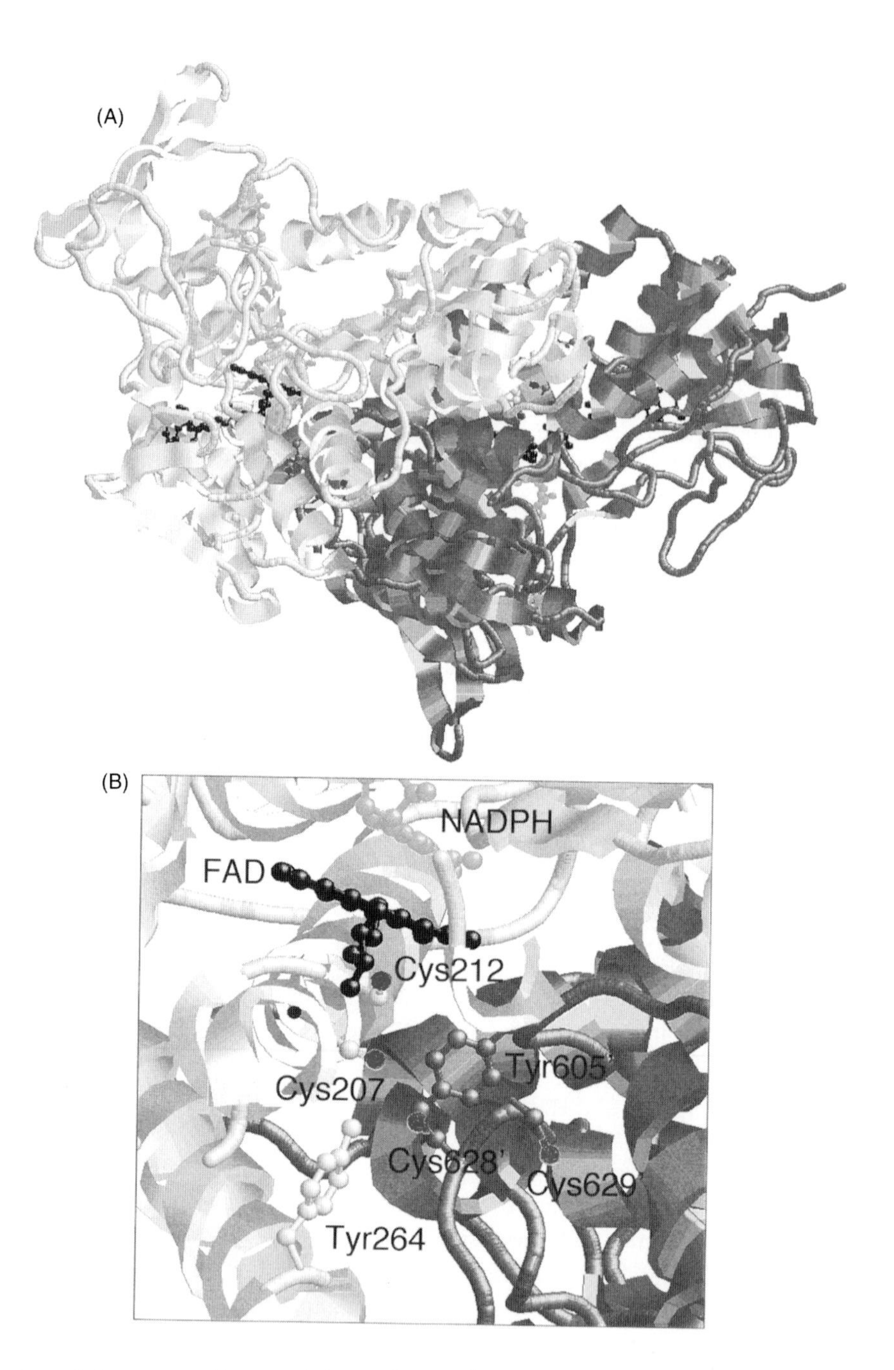

Figure 5. Structure of mercuric ion reductase homodimer from *Bacillus* RC607 [21]
(**A**) Light and dark chains are individual polypeptides. (**B**) Close-up of the left-hand active site in view (**A**). Black structure is FAD viewed end-on, light blue is NADPH, and dark blue dots are cysteine sulphurs. Primes indicate residues from the other monomer at the interface. Cys207 and Cys212 are the redox-active pair and Cys628′ and Cys629′ are the C-terminal pair. Structures were generated using RasMac v2.6.

the reduction potential to a level that would inhibit reduction of the bound Hg(II). Thus a key mechanistic question is how the enzyme utilizes its cysteines to achieve both jobs of trapping and reducing Hg(II). *In vivo*, all four cysteines are essential, since mutation of any of them to alanine in the context of the full *mer* operon yields a physiologically deficient enzyme [25,26]. *In vitro*, the defects vary and, in some cases, are dependent on the conditions used to evaluate the mutant. The clearest result is that mutation of either of the redox-active cysteines results in complete loss of rapid Hg(II) reduction under any conditions, indicating that both residues are essential for electron transfer to Hg(II) at catalytically relevant rates, and may be required for formation of the reducible Hg(II) complex [25]. Confirmation of the latter conclusion comes from recent results with the double Cys→Ala mutant of the C-terminal cysteines. This mutant is inactive in steady-state assays where the substrate is a Hg–dithiol complex, as in the cell [26]. However, rapid-reaction studies monitoring the spectral properties of the flavin cofactor show that when the Hg(II) substrate is a complex with smaller and much weaker ligands than thiols (e.g. $HgBr_2$) [27], the enzyme rapidly forms a Hg(II) complex with the redox-active cysteines that is then reduced at a rate faster than catalytic turnover. Current studies of this mutant using Hg–dithiol complexes indicate a more limited access of these complexes to the active site and, perhaps, an incomplete displacement of the non-enzymic thiol ligands. Comparison with the wild-type enzyme and the single Cys→Ala mutants of the C-terminal cysteines [28] suggests that both C-terminal cysteines are essential for fully displacing the non-enzymic thiol ligands before Hg(II) is transferred to the redox-active cysteines

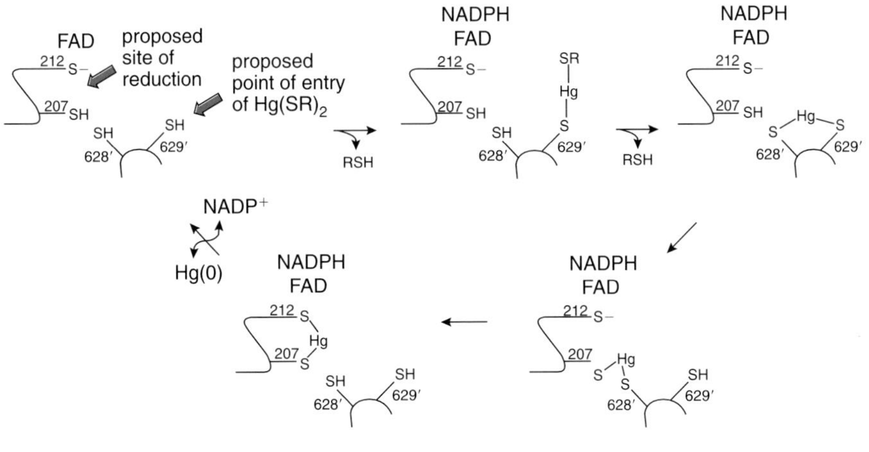

Figure 6. Proposed pathway for transfer of Hg(II) from a $Hg(SR)_2$ substrate
Transfer is via the C-terminal cysteines into a complex with the redox-active cysteine pair, adjacent to the flavin, where reduction can occur. Adapted with permission from [29]. ©(1999) American Chemical Society.

for reduction [29]. Thus, our current view is that the predominant role of the C-terminal cysteines is to provide the trapping site for Hg(II), whereas the redox-active cysteines provide the reducible complex of Hg(II) (Figure 6).

The story, however, is not quite complete there. Extensive evidence indicates that the two active sites in the MerA homodimer exhibit different thermodynamic and kinetic properties in many reactions of ligands with the wild-type and mutant forms of the enzyme. These observations have led to the proposal that the enzyme may function by an alternating-sites mechanism where the properties of the monomers on each dimer are always asymmetric, but alternate in a coupled fashion throughout the reaction [30]. Such a mechanism may facilitate catalysis when there is a need for conversion from tight binding to weak binding of a ligand during the catalytic cycle; the best-documented example of an alternating-sites mechanism is in ATP synthases [31]. Although more difficult to demonstrate in mercuric ion reductase, studies are currently underway to evaluate the role of site–site interactions in the catalytic mechanism of this enzyme.

Perspectives

Although risky, the strategy of importing toxic Hg(II) to eliminate it is worth the risk because the unique chemical properties of Hg(0) more effectively remove it from the environment of the organism than complexation or simple exportation. However, the risk demands fine tuning of the pathway proteins for the most efficient detection and handing-off to the final detoxification protein. At present, the exquisite design for detection and regulation by MerR is the best-understood part of the pathway. Although some important features have been identified for each of the other proteins, much work remains to develop a clear picture of how successive proteins in the path function and interact with each other to accomplish the overall goal of detoxification.

Summary

- *The most common bacterial mechanism for resistance to mercuric-ion species involves intracellular reduction of Hg(II) to Hg(0).*
- *Key proteins of the pathway typically include: MerR, which regulates pathway expression; MerP, which protects the external environment; MerT or MerC, which transport Hg(II) species across the inner membrane; MerA, which catalyses reduction of Hg(II); and sometimes MerB, which catalyses cleavage of C–Hg bonds in organomercurials.*
- *Cysteine residues of varying number are arranged in each of the key proteins to optimize their unique roles in sensing (high affinity), transporting (exchangeability), and reducing (redox accessibility) Hg(II).*

- *Nature's regulator of this pathway, MerR, is an exquisitely sensitive, Hg(II)-binding, DNA-binding protein that holds the system primed for immediate transcription at the slightest influx of Hg(II).*

Our current studies on the mechanism of mercuric ion reductase are supported by a grant from the United States National Institutes of Health NIGMS (GM50670). We thank Dr. Emil Pai for sharing the co-ordinates of mercuric ion reductase.

References

1. Silver, S. & Phung, L.T. (1996) Bacterial heavy metal resistance: new surprises. *Annu. Rev. Microbiol.* **50**, 753–789
2. Osborn, A.M., Bruce, K.D., Strike, P. & Ritchie, D.A. (1997) Distribution, diversity and evolution of the bacterial mercury resistance (*mer*) operon. *FEMS Microbiol. Rev.* **19**, 239–262
3. Klein, J., Altenbuchner, J. & Mattes, R. (1997) Genetically modified *Escherichia coli* for colorimetric detection of inorganic and organic compounds. *Exs* **80**, 133–151
4. Summers, A.O. (1986) Organization, expression, and evolution of genes for mercury resistance. *Annu. Rev. Microbiol.* **40**, 607–634
5. Summers, A.O. (1992) Untwist and shout: a heavy metal-responsive transcriptional regulator. *J. Biol. Chem.* **174**, 3097–3101
6. Ansari, A.Z., Bradner, J.E. & O'Halloran, T.V. (1995) DNA-bend modulation in a repressor-to-activator switching mechanism. *Nature (London)* **374**, 371–375
7. Ralston, D.M. & O'Halloran, T.V. (1990) Ultrasensitivity and heavy-metal selectivity of the allosterically modulated MerR transcription complex. *Proc. Natl. Acad. Sci. U.S.A.* **87**, 3846–3850
8. Watton, S.P., Wright, J.G., MacDonnell, F.M., Bryson, J.W., Sabat, M. & O'Halloran, T.V. (1990) Trigonal mercuric complex of an aliphatic thiolate: a spectroscopic and structural model for the receptor site in the Hg(II) biosensor merR. *J. Am. Chem. Soc.* **112**, 2824–2826
9. Utschig, L.M., Bryson, J.W. & O'Halloran, T.V. (1995) Mercury-199 NMR of the metal receptor site in merR and its protein-DNA complex. *Science* **268**, 380–385
10. Helmann, J.D., Ballard, B.T. & Walsh, C.T. (1990) The merR metalloregulatory protein binds mercuric ion as a tricoordinate, metal-bridged dimer. *Science* **247**, 946–948
11. Hamlett, N.V., Landale, E.C., Davis, B.H. & Summers, A.O. (1992) Roles of the Tn21 merT, merP and merC gene products in mercury resistance and mercury binding. *J. Bacteriol.* **174**, 6377–6385
12. Morby, A.P., Hobman, J.L. & Brown, N.L. (1995) The role of cysteine residues in the transport of mercuric ions by the Tn501 merT and merP mercury-resistance proteins. *Mol. Microbiol.* **17**, 25–35
13. Sahlman, L. & Skärfstad, E.G. (1993) Mercuric ion binding abilities of merP variants containing only one cysteine. *Biochem. Biophys. Res. Commun.* **196**, 583–588
14. Eriksson, P.-O. & Sahlman, L. (1993) ^{1}H NMR studies of the mercuric ion binding protein merP: sequential assignment, secondary structure and global fold of oxidized merP. *J. Biomol. NMR* **3**, 613–626
15. Steele, R.A. & Opella, S.J. (1997) Structures of the reduced and mercury-bound forms of merP, the periplasmic protein from the bacterial mercury detoxification system. *Biochemistry* **36**, 6885–6895
16. Uno, Y., Kiyono, M., Tezuka, T. & Pan-Hou, H. (1997) Phenylmercury transport mediated by *merT-merP* genes of *Pseudomonas* K-62 plasmid pMR26. *Biol. Pharm. Bull.* **20**, 107–109
17. Begley, T.P., Walts, A.E. & Walsh, C.T. (1986) Mechanistic studies of a protonolytic organomercurial cleaving enzyme: bacterial organomercurial lyase. *Biochemistry* **25**, 7192–7200

18. Moore, M.J., Distefano, M.D., Zydowsky, L.D., Cummings, R.T. & Walsh, C.T. (1990) Organomercurial lyase and mercuric ion reductase: nature's mercury detoxification catalysts. *Acc. Chem. Res.* **23**, 301–308
19. Schottel, J.L. (1978) The mercuric and organomercurial detoxifying enzymes from a plasmid-bearing strain of *Escherichia coli*. *J. Biol. Chem.* **253**, 4341–4349
20. Distefano, M.D., Moore, M.J. & Walsh, C.T. (1990) Active site of mercuric reductase resides at the subunit interface and requires Cys_{135} and Cys_{140} from one subunit and Cys_{558} and Cys_{559} from the adjacent subunit: evidence from *in vivo* and *in vitro* heterodimer formation. *Biochemistry* **29**, 2703–2713
21. Schiering, N., Kabsch, W., Moore, M.J., Distefano, M.D., Walsh, C.T. & Pai, E.F. (1991) Structure of the detoxification catalyst mercuric ion reductase from *Bacillus* sp. strain RC607. *Nature (London)* **352**, 168–172
22. Brown, N.L., Ford, S.J., Pridmore, R.D. & Fritzinger, D.C. (1983) Nucleotide sequence of a gene from the *Pseudomonas* transposon Tn*501* encoding mercuric reductase. *Biochemistry* **22**, 4089–4095
23. Miller, S.M., Ballou, D.P., Massey, V., Williams, Jr., C.H. & Walsh, C.T. (1986) Two-electron reduced mercuric reductase binds Hg(II) to the active site dithiol but does not catalyse Hg(II) reduction. *J. Biol. Chem.* **261**, 8081–8084
24. Miller, S.M., Moore, M.J., Massey, V., Williams, Jr., C.H., Distefano, M.D., Ballou, D.P. & Walsh, C.T. (1989) Evidence for the participation of Cys_{558} and Cys_{559} at the active site of mercuric reductase. *Biochemistry* **28**, 1194–1205
25. Distefano, M.D., Au, K.G. & Walsh, C.T. (1989) Mutagenesis of the redox-active disulphide in mercuric ion reductase: catalysis by mutant enzymes restricted to flavin redox chemistry. *Biochemistry* **28**, 1168–1183
26. Moore, M.J. & Walsh, C.T. (1989) Mutagenesis of the N- and C-terminal cysteine pairs of Tn501 mercuric ion reductase: consequences for bacterial detoxification of mercurials. *Biochemistry* **28**, 1183–1194
27. Engst, S. & Miller, S.M. (1998) Rapid reduction of Hg(II) by mercuric ion reductase does not require the conserved C-terminal cysteine pair using $HgBr_2$ as substrate. *Biochemistry* **37**, 11496–11507
28. Moore, M.J., Miller, S.M. & Walsh, C.T. (1992) C-terminal cysteines of Tn*501* mercuric ion reductase. *Biochemistry* **31**, 1677–1685
29. Engst, S. & Miller, S. (1999) Alternative routes for entry of HgX_2 into the active site of mercuric ion reductase depend on the nature of the X ligands. *Biochemistry* **38**, 3519–3529
30. Miller, S.M., Massey, V., Williams, Jr., C.H., Ballou, D.P. & Walsh, C.T. (1991) Communication between the active sites in dimeric mercuric ion reductase: an alternating sites hypothesis for catalysis. *Biochemistry* **30**, 2600–2612
31. Boyer, P.D. (1989) A perspective of the binding change mechanism for ATP synthesis. *FASEB J.* **3**, 2164–2178

3

Non-haem iron-containing oxygenases involved in the microbial biodegradation of aromatic hydrocarbons

Eric D. Coulter[1] and David P. Ballou[2]

Department of Biological Chemistry, University of Michigan, Ann Arbor, MI 48109-0606, U.S.A.

Introduction

The accumulation of potentially toxic hydrocarbons in our soil and water systems is a growing problem. Many hydrocarbons enter the environment as a result of the biodegradation of natural materials such as lignin, a complex polymer component of plant cell walls. However, others, such as industrial waste products, residues from pesticides and herbicides, and emissions from the normal activities of our modern industrial civilization, are of particular concern because they are often toxic and/or carcinogenic, and many are highly stable and resistant to degradation. We are fortunate that micro-organisms can metabolize a wide variety of these aromatic substances for use as sources of carbon and energy. It is hoped that the study of these microbial degradative pathways will lead to the development of processes for degrading some of the more recalcitrant compounds, and also will help us to learn which types of compound that we might use in domestic and industrial processes will be amenable to biodegradation.

[1]*Present address: Department of Chemistry, University of Georgia, Athens, GA, U.S.A*

[2]*To whom correspondence should be addressed.*

Similarities and diversities among the degradative pathways for several common aromatic compounds are highlighted in Figure 1. Significant in microbial metabolism is the presence of both aerobic and anaerobic pathways for the degradation of the same compound, as is shown for benzoate and toluene. Toluene, in particular, exemplifies the variety of metabolic routes by which a single compound can be utilized as a carbon source. Elucidation of anaerobic pathways is in the early stages, but one common pathway (Figure 1, below the dotted line) involves conversion to benzoyl-CoA, followed by aromatic ring reduction and cleavage to yield acetyl-CoA [1]. Aerobic pathways are much better understood (Figure 1, above the dotted line), and generally proceed via initial O_2 incorporation to the benzene ring to form catechols, which are subsequently cleaved in a second O_2-dependent reaction (see below and Chapter 11 in this volume). Regardless of the pathway followed, aromatic compounds are converted to aliphatic products that enter into intermediary metabolism for use in biosynthetic pathways or for the generation of energy. An excellent overview and source of references of major hydrocarbon

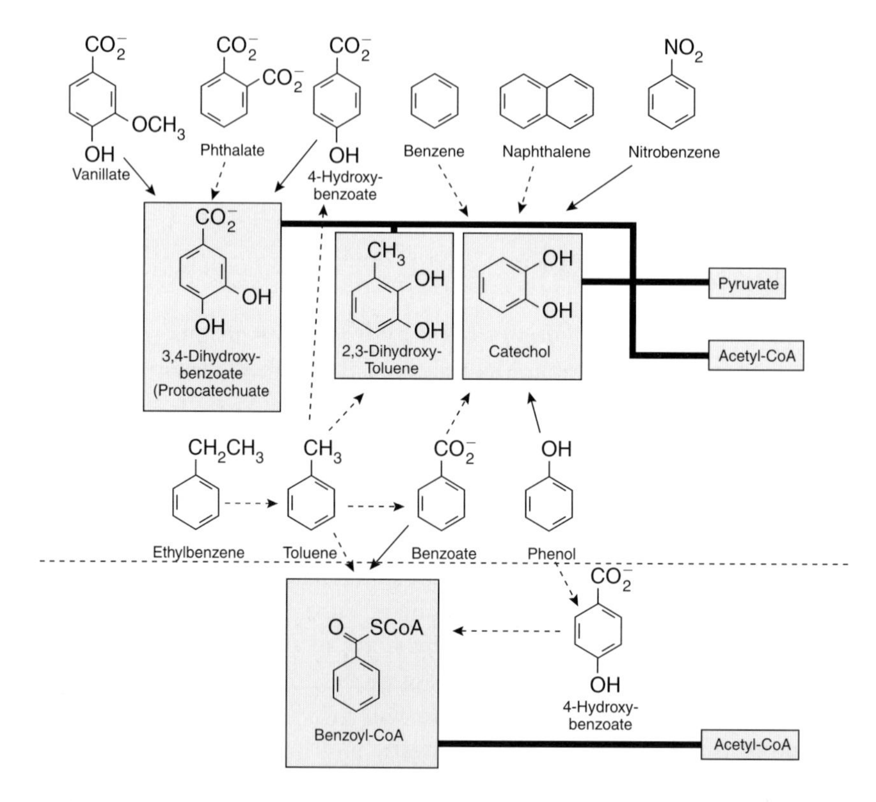

Figure 1. Biodegradative pathways for several common aromatic compounds
Aerobic pathways are indicated above the dashed line, anaerobic pathways below the dashed line. Solid arrows show one-step reactions and dashed arrows indicate two or more steps. Intermediates common to several pathways are shown in boxes. Heavy lines indicate major metabolic products that result from the given aromatic compounds. These products participate as energy sources in the hosts.

biodegradation pathways, as well as links to many biochemically useful sites, is provided by the University of Minnesota Biocatalysis/Biodegradation Database, which can be accessed on the World Wide Web at http://dragon.labmed.umn.edu/~lynda/index.html.

Incorporation of O_2 into the aromatic rings, as catalysed by oxygenases, is central to the aerobic degradation of these compounds. Although the oxidation of organic compounds is thermodynamically favourable (note that most burn very well), in the absence of a catalyst the direct reaction between O_2 and most organic compounds is so slow that it is of little biological significance. Oxygenases overcome the kinetic barriers associated with the low reactivity of O_2 by either making O_2 more reactive (O_2 activation) or by making the substrate more acceptable to attack by O_2 (substrate activation). Some examples and hypotheses of how this occurs are presented in this article as well as in Chapters 4, 5 and 11 in this volume.

Both aromatic-ring-oxygenation and aromatic-ring-cleaving reactions are catalysed by oxygenases, which can be divided into two general categories according to the number of oxygen atoms that are inserted into the substrate. Mono-oxygenases incorporate one atom of molecular oxygen (O_2) into substrate (S), whereas the other is reduced to water (reaction 1). Dioxygenases incorporate both oxygen atoms (intramolecular; reaction 2) into substrate or one atom (intermolecular; reaction 3) into substrate and the other into an associated cofactor (R). In addition to this review, other Chapters in this volume on cytochrome P450 and catechol oxygenases also deal with mono-oxygenases and dioxygenases as shown in reactions (1) and (2) below.

$$S + O_2 \xrightarrow{\text{mono-oxygenase}} SOH + H_2O \quad (1)$$

$$S + O_2 \xrightarrow{\text{dioxygenase}} S(OH)_2 \quad (2)$$

$$S + R + O_2 \xrightarrow{\text{dioxygenase}} SOH + RO \quad (3)$$

Oxygen activation at non-haem iron centres

Oxygenases often employ a transition metal, in the form of a mononuclear, binuclear or haem centre, or an organic cofactor, such as flavin, to facilitate O_2 activation. Iron and copper are the most commonly used metals because, in their lower oxidation states, these metals can form complexes with O_2, substrate or both, and affect the electronic structure of the bound moiety to alter its reactivity. Complete reduction of O_2 to water proceeds by four sequential one-electron steps, as shown in the top portion of Figure 2. Of greatest significance to biological modes of O_2 activation are the one- and two-electron reductions to superoxide and peroxide, respectively. Reduction of O_2 in the presence of iron is typical of metal-assisted reactions (Figure 2). Iron in the ferrous [Fe(II)] oxidation state can react directly with O_2, forming an

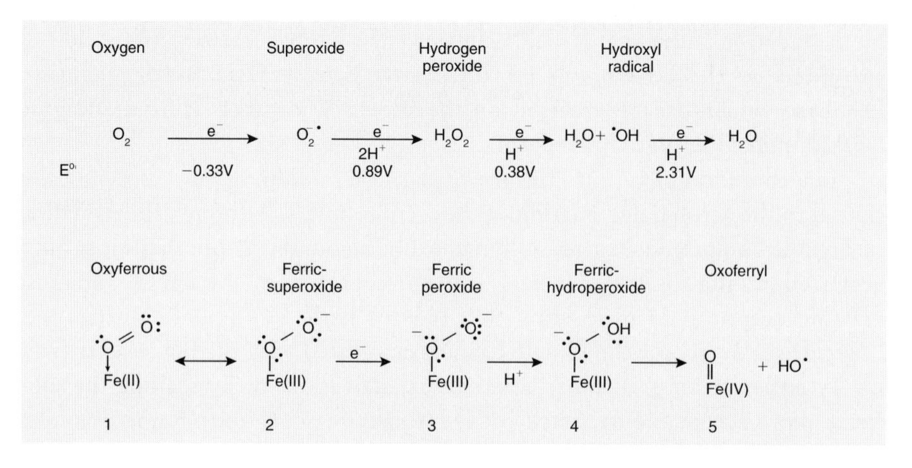

Figure 2. Successive one-electron reductions of molecular oxygen (O_2) to water
(Top) Free O_2 in aqueous solution shown with the redox potentials at pH 7. (Bottom) Iron-dependent activation of O_2. Conversion of **2** to **4** may occur as shown or perhaps, *in vivo*, with the protonation step first, followed by reduction.

oxyferrous complex **1** (see Figure 2). Reduction of the metal-bound O_2 to superoxide does not require input of additional electrons; instead, delocalization of electron density from Fe(II) to the bound O_2 (analogous to a one-electron reduction) results in a ferric–superoxide complex **2**. Reduction of species **2** yields a ferric–peroxide complex **3**, and protonation forms a ferric–hydroperoxide complex **4**. Metal–peroxo species can be end-on as shown, or side-on as discussed later. Although species **2**, **3** or **4** are potential candidates for oxygenating substrates, the oxygenation of unreactive hydrocarbons or aromatic compounds is usually thought to require a high-valency oxo-species **5**, formed by the cleavage of the O–O bond in species **4**. Either heterolytic or homolytic cleavage can occur, depending on the nature of the site; homolytic cleavage is shown in Figure 2. This process is discussed more fully in Chapters 4 and 5 in this volume on peroxidases and cytochrome P450.

The reduction of O_2 requires a source of electrons. Oxygenases can accept electrons not only from external reductants such as reduced pyridine nucleotides [e.g. mononuclear Fe(II) aromatic-ring dioxygenases], but also from the substrate itself (e.g. catechol dioxygenases), or an associated organic cofactor such as a pterin (e.g. aromatic amino acid hydroxylases). Examples of four major classes of non-haem iron-containing enzymes involved in aromatic oxygenations are shown along with their sources of electrons in Table 1. These non-haem iron oxygenases exhibit rich diversity in structure and mechanism, each employing several of the intricacies assisting O_2 incorporation described above.

Catechol dioxygenases

Catechol oxygenases incorporate both atoms of a single O_2 molecule into a catechol substrate [2]. Figure 1 shows how the formation of catechols is

Table 1. Major classes of non-haem iron oxygenase that catalyse aromatic-ring oxygenations

Class	Electron source	Examples
I. Catechol dioxygenases		
Intradiol	Substrate	Catechol 1,2-dioxygenase Protocatechuate dioxygenase
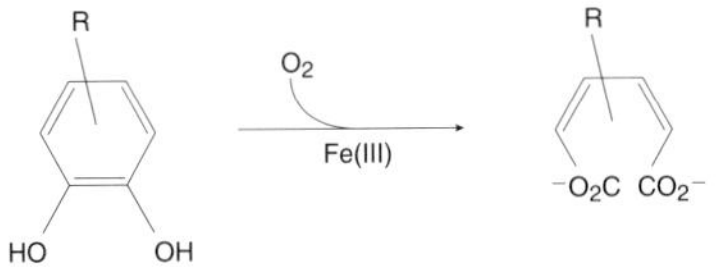		
Extradiol	Substrate	Catechol 2,3-dioxygenase 2,3-Dihydroxybiphenyl 1,2-dioxygenase
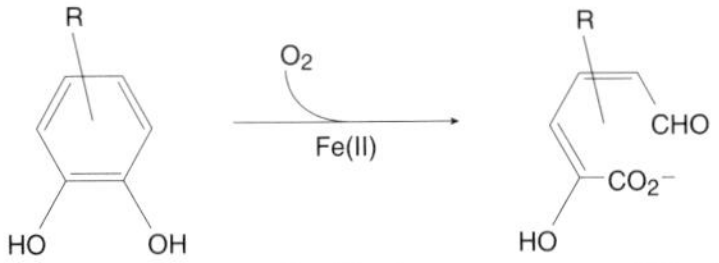		
II. Mononuclear Fe(II) dioxygenases (Rieske oxygenases)		
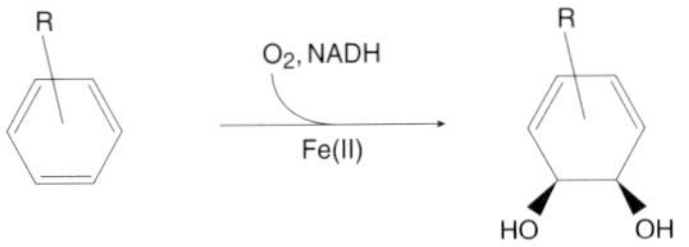	NADH	Benzene 1,2-dioxygenase Toluene 1,2-dioxygenase Phthalate 3,4-dioxygenase
III. Fe(II)–pterin-dependent oxygenases		
Other aromatic hydroxylases	Tetrahydropterin	Benzoate 4-hydroxylase Mandelate 4-hydroxylase
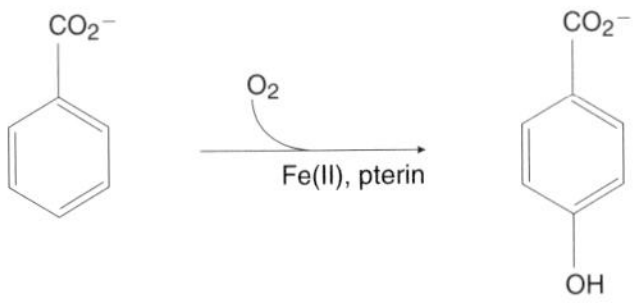		
Amino acid hydroxylases	Tetrahydropterin	Tyrosine hydroxylases Phenylalanine hydroxylase Tryptophan hydroxylase
IV. Dinuclear iron mono-oxygenases		
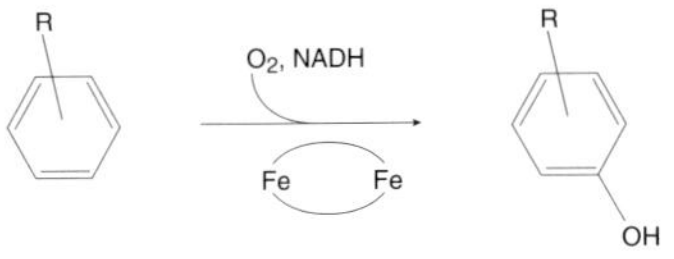	NADH	Toluene 2-mono-oxygenase Toluene 3-mono-oxygenase Toluene 4-mono-oxygenase Xylene mono-oxygenase Phenol mono-oxygenase

common in the biodegradation of aromatic compounds. The action of the catechol dioxygenases results in fission of the aromatic ring to yield aliphatic products. This catecholic ring cleavage can proceed via either of two routes: (i) intradiol dioxygenases employ Fe(III) to break the C–C bond between the hydroxy groups; or (ii) extradiol dioxygenases employ Fe(II) to break the C–C bond adjacent to the hydroxy groups; see Table 1 for the general reactions.

At first glance the oxidative C–C bond-cleavage reactions catalysed by the extradiol and intradiol catechol dioxygenases may appear similar, but studies

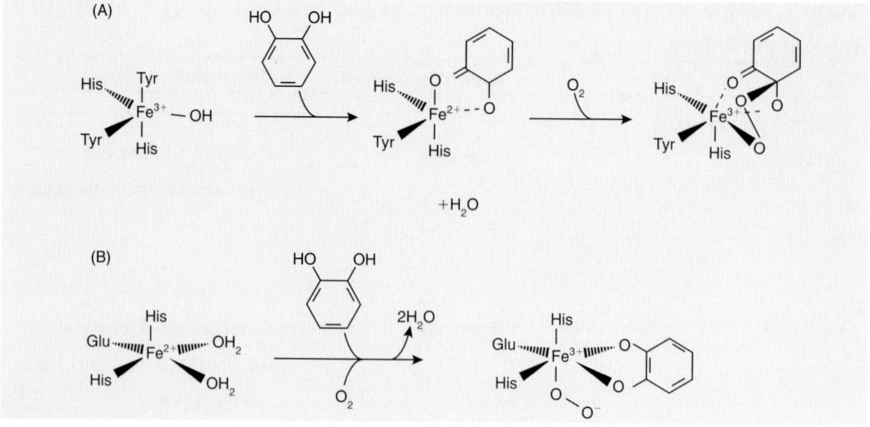

Figure 3. Binding modes of catechols and oxygen to intradiol and extradiol dioxygenases

(**A**) Substrate activation and oxygenation by an intradiol catechol dioxygenase. Note that the substrate binds to the Fe(III) centre and becomes partially semiquinone in character, allowing direct attack on the substrate by O_2. (**B**) Binding of substrate and O_2 by an extradiol catechol dioxygenase. Here the Fe(II)-bound O_2 attains superoxide character, allowing direct attack on the bound substrate.

have shown the mechanisms to be quite different [2]. In the case of the intradiol dioxygenases, Fe(III) is proposed to co-ordinate to the substrate and activate it for attack by O_2 (Figure 3A). In contrast, in the extradiol dioxygenases, both substrate and O_2 co-ordinate the Fe(II) site concurrently (Figure 3B), and it is the O_2 that is activated to attack the benzene ring. External sources of electrons are not required for either intradiol or extradiol mechanisms because the substrate supplies both electrons. Catechol dioxygenases are described in more detail in Chapter 11 in this volume and will not be discussed further here.

Mononuclear Fe(II) dioxygenases (Rieske oxygenases)

Dihydroxylation is commonly the first step in the aerobic microbial degradation of unactivated benzene rings. The resulting non-aromatic product (a *cis*-diol) is converted by a dehydrogenase to a catechol with concomitant reduction of NAD to NADH; a catechol dioxygenase then cleaves the ring as described above. Benzene, benzoate, toluene, naphthalene and phthalate are representative of a variety of compounds (more than 40 such systems have been identified) that can be metabolized in this way. When the reaction occurs with a halogenated aromatic compound, the resulting diol often re-aromatizes by eliminating HX (X=F, Cl or Br), which contributes to the detoxification process. Oxygenation occurs at a mononuclear Fe(II) site that has structural and sequence similarity with several enzymes recognized recently as a family [2] of oxygenases, as discussed in the following sections. All of the Fe(II)-containing aromatic *cis*-diol-forming dioxygenases are multicomponent systems that, in addition to an oxygenase, require one or two electron-transfer

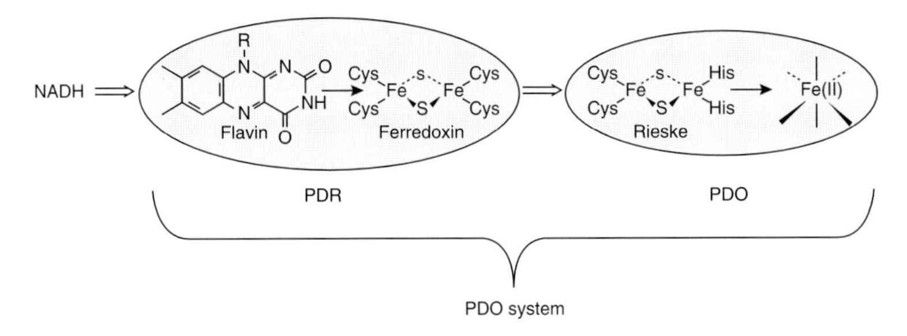

Figure 4. Protein components of the class-I PDO system

proteins to shuttle electrons from NADH to the site of oxygenation [3]. These electron-transfer proteins utilize flavins (flavins can accept a hydride from NADH and transfer the two electrons, one at a time, using the flavin semiquinone as an intermediate) and [2Fe–2S] centres to shuttle electrons from the flavin to the site of oxygenation. The *cis*-diol-forming dioxygenases are classified according to the number and cofactor requirements of the associated electron-transfer proteins (Table 2).

The phthalate dioxygenase (PDO) system from *Bulkholderia (Pseudomonas) cepacia* is a class-I system (the simplest type) that has been characterized extensively and has served as a spectroscopic and mechanistic model for other *cis*-diol-forming dioxygenases [4]. This two-component system consists of a 36-kDa FMN- and ferredoxin-containing reductase (phthalate dioxygenase reductase or PDR) [5] that serves to transfer electrons from NADH, and a 200-kDa homotetrameric oxygenase (PDO) that contains a Rieske [2Fe–2S] centre [6] and a mononuclear Fe(II) site (Figure 4). Thus, each subunit of PDO contains three irons, two in an [2Fe–2S] cluster, and a third at a distinct site proposed to be the site of substrate oxygenation. The [2Fe–2S] clusters of this and other oxygenase components of this class exhibit higher redox potentials and considerably different spectroscopic properties than the clusters of typical plant-type ferredoxins, such as that present in PDR [5]. Instead, they closely resemble the Rieske centres found in the bc_1 components of respiratory chains, and thus these oxygenases are often referred to by the trivial name, Rieske oxygenases. The different chemical and spectroscopic

Table 2. Classification of Fe(II)-containing *cis*-dihydrodiol dioxygenases

Class	Components	Example
(I)	{FMN or (FAD), [2Fe–2S]} → $[2Fe–2S]_{Rieske}$–Fe(II) Reductase → Oxygenase	Phthalate dioxygenase Benzoate dioxygenase
(II)	FAD → [2Fe–2S] → $[2Fe–2S]_{Rieske}$–Fe(II) Reductase → Ferredoxin → Oxygenase	Benzene 1,2-dioxygenase
(III)	FAD, [2Fe–2S] → $[2Fe–2S]_{Rieske}$ → $[2Fe-2S]_{Rieske}$–Fe(II) Reductase → Electron-transfer protein → Oxygenase	Naphthalene dioxygenase

properties of the Rieske centre largely result from the substitution of two histidines for two cysteines at one iron of the [2Fe–2S] cluster [6]. The very recent elucidation of a crystal structure for the related Rieske oxygenase, naphthalene 1,2-dioxygenase (NDO), confirms this assignment of the [2Fe–2S] ligands [7].

The PDO system catalyses the first step in the breakdown of phthalate [8], forming exclusively *cis*-4,5-dihydroxydihydrophthalate (see reaction 4). This non-aromatic product is converted in a two-step process to protocatechuate, a common intermediate of several biodegradation pathways (Figure 1). Protocatechuate is cleaved to an aliphatic product by protocatechuate dioxygenase; see Chapter 11 in this volume.

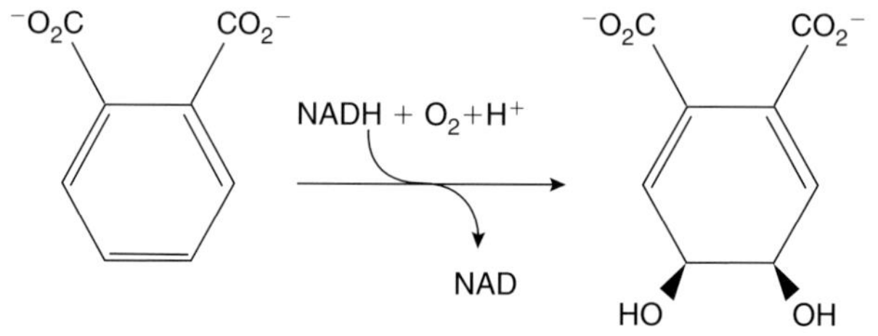

$$NADH + O_2 + H^+ \rightarrow NAD \qquad (4)$$

Structure of the mononuclear iron centre

The mononuclear Fe(II) is essential for catalytic activity [4] and is proposed to be the site of O_2 binding, activation and substrate hydroxylation. EXAFS (extended X-ray absorption fine structure spectroscopy) measurements suggest that Fe(II) is ligated exclusively by nitrogen and O_2 donor ligands [9]. Crystal structures have recently been determined for five other non-haem Fe(II)-containing enzymes that catalyse O_2-dependent reactions, including the extradiol catechol dioxygenase, 2,3-biphenyl 1,2-dioxygenase [10], lipoxygenases from rabbit [11] and soya bean [12,13], isopenicillin *N*-synthase

Lipoxygenase (rabbit)

Lipoxygenase (soya bean)

Isopenicillin *N*-synthase

Phenylalanine hydroxylase

2,3-Dihydroxybiphenyl 1,2-dioxygenase

Naphthalene dioxygenase

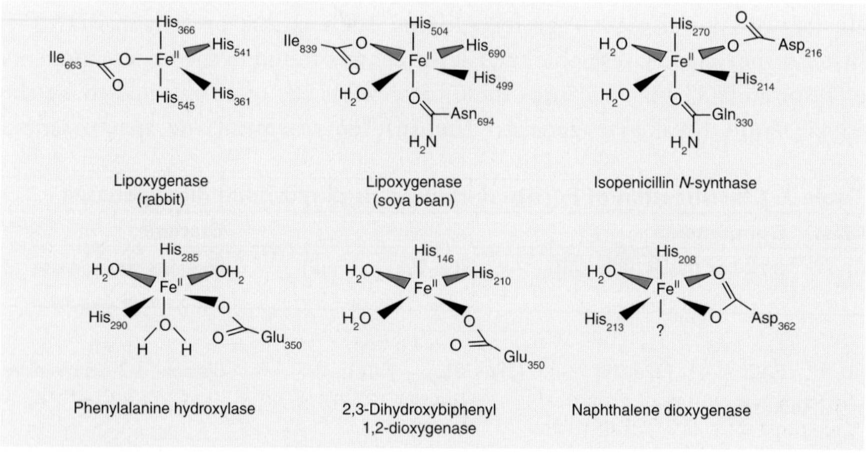

Figure 5. Structures of the mononuclear Fe(II) sites of six crystallographically determined O_2-utilizing enzymes [10–16]

[14], and tyrosine and phenylalanine hydroxylases [15,16]. In each of these structures the mononuclear Fe(II) site has a mixture of histidyl, carboxylate and solvent-derived water ligands. Sequence alignments of several *cis*-diol-forming dioxygenases identified a highly conserved region ($\mathbf{W}X_{4-5}\mathbf{E}X_{3-4}\mathbf{D}X_2\mathbf{H}X_{4-5}\mathbf{H}$; where bold indicates residues nearly always found in these sequences) containing two potential carboxylate donors and two histidines that were proposed to be mononuclear site ligands [3,17]. Site-directed mutagenesis of each of the equivalent residues by alanine in the related toluene 1,2-dioxygenase enzyme resulted in total loss of activity, suggesting that these residues were ligands to Fe(II) [18]. Even with this information, largely because of the inadequacy of most spectroscopic techniques for investigating Fe(II), a detailed structure of the mononuclear site proved elusive for many years. Only now, with the crystal structure for NDO [7], can we begin to assign ligands to the mononuclear metal site with some degree of certainty. The mononuclear site of NDO is co-ordinated by an amino acid motif ($\mathbf{H}X_4\mathbf{H}X_{148}\mathbf{D}$) and a single water molecule (Figure 5). Sequence alignment identifies a similar motif ($\mathbf{H}X_{4-5}\mathbf{H}X_{138-157}\mathbf{D}$) in several Rieske dioxygenases, but the exact assignment, especially of the final aspartate (or glutamate), is uncertain for most of these oxygenases. The NDO structure

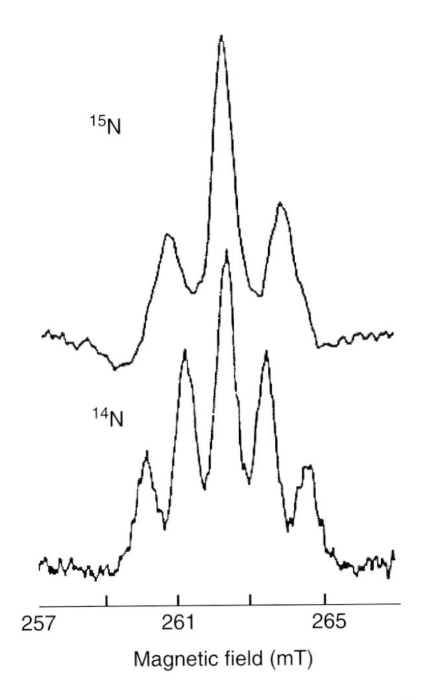

Figure 6. Low-field super-hyperfine splitting of Cu(II)-reconstituted PDO in which all nitrogen atoms are ^{14}N (bottom) or ^{15}N (top)

EPR spectra of $^{63}Cu(II)$-reconstituted PDO (200 μM monomer) were recorded at X-band at 120 K. EPR conditions: microwave frequency, ≈9.1717 MHz; microwave power, 2 mW; modulation amplitude, 5.0 G.

also shows that the glutamate in the initially recognized sequence is not very near the iron, but may be important in subunit interactions, whereas the aspartate, although not a ligand, hydrogen bonds to the first ligating histidine [7]. Nevertheless, both analogous residues are critical to the function of toluene dioxygenase. This shows how it can be folly to assign ligands to a metal based solely on sequence analysis.

Fe(II) can readily be removed from the mononuclear site of PDO and replaced by various divalent metals, including Cu(II) [4], which can be used as a spectroscopic probe for investigating ligand structures. Although the resulting protein is catalytically inactive, it is structurally very similar to native PDO, implying that the copper binding is similar to that of Fe(II). EPR spectroscopy has been used to examine various preparations of PDO reconstituted with ^{63}Cu(II) [19]. The 3/2 nuclear spin (I=3/2) of Cu(II) gives rise to EPR signals with four well-defined hyperfine lines ($2I+1=4$) on both the g-parallel (low-field) and the g-perpendicular (high-field) EPR absorbance bands. These copper hyperfine lines can be split further by interactions with ligands that contain nuclei with nuclear spins, such as ^{14}N (I=1), which gives super-hyperfine resolution. Each of the four g-parallel lines of Cu(II)PDO can be clearly resolved into five components of relative intensities; 1:2:3:2:1 (Figure 6 shows the super-hyperfine splittings of one of the g-parallel lines). These five bands are consistent with the presence of two nitrogenous ligands ($2nI+1=5$), where $n=2$. When ^{15}N (I=1/2) is present in all amino acids [the bacteria were grown with $(^{15}NH_4)_2SO_4$ as its nitrogen source], as predicted, the copper super-hyperfine for Cu(II)PDO has only three peaks with relative intensities 1:2:1 ($2nI+1=3$, $n=2$), as shown in Figure 6. Cu(II)PDO in which only the nitrogens of histidine have been uniformly labelled with ^{15}N also exhibited three components (1:2:1). These results provide direct evidence for the ligation of two histidines to copper bound at the mononuclear site.

Mechanism

In addition to *cis*-dihydroxylation reactions, Rieske oxygenases can catalyse mono-oxygenase reactions; for example, naphthalene 1,2-dioxygenase catalyses the conversion of indane to indanol [20] and putidamonooxin primarily functions as an *o*-demethylase [21]. These observations suggest that a common form of active O_2 may be responsible for both mono- and dioxygenase reactions, but that the type of oxygenation is determined by the nature of the substrate and how it is bound at the active site. The haem-containing family of enzymes, collectively called the cytochromes P450, are considered the prototypical mono-oxygenases (see Chapter 5 in this volume). They catalyse a diversity of reactions, including hydroxylation of aromatic compounds and *o*-demethylations. Rieske oxygenases and cytochrome P450 both require NAD(P)H, O_2 and substrate for activity, and in the presence of

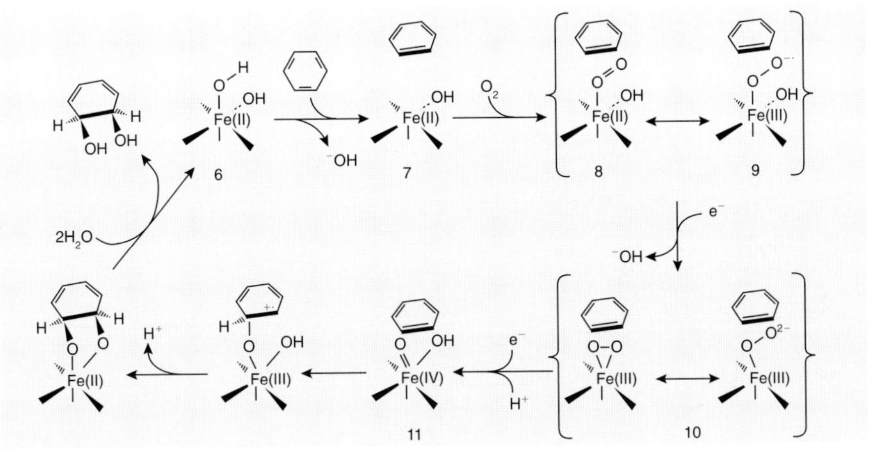

Figure 7. Proposed reaction cycle for the mononuclear Fe(II)-containing aromatic-ring hydroxylases

substrates that are mono-oxygenated, exhibit identical reaction stoichiometries:

$$1 \text{ mol NAD(P)H} + 1 \text{ mol } O_2 \longrightarrow 1 \text{ mol product} + 1 \text{ mol } H_2O + 1\text{mol NAD(P)}$$

Considering the similarities in both the reactions catalysed and the overall stoichiometries, a mechanism similar to that widely accepted for the cytochrome P450 has been proposed for the Rieske oxygenases; see Figure 7. The cycle consists of three individual stages: substrate binding, O_2 activation and substrate oxygenation. Convincing evidence in support of an initial decrease in co-ordination number of the 6-co-ordinate Fe(II)-resting state **6** (see Figure 7) to a 5-co-ordinate substrate-bound form **7** has been presented [9,22]. The loss of a metal-co-ordinating ligand, most likely H_2O or OH^-, opens up a site for O_2 to bind (oxyferrous; **8**). As shown in Figure 2, electron density can shift from Fe(II) to O_2, generating ferric–superoxide species **9**. Transfer of an electron from NADH via PDR and the Rieske [2Fe–2S] centre (Figure 3) reduces the bound superoxide to peroxide **10**. The binding may be either side-on or end-on as shown. The hydroxylation of non-optimal substrates results in deviation from the stoichiometry shown above. Instead, the amount of hydroxylated product decreases and hydrogen peroxide is produced, presumably from the protonation of species **10**. In catalysis, cleavage of the O–O in species **10** generates a strongly oxidizing high-valency oxoferryl [Fe(IV)=O] species **11**, analogous to the proposed oxygenating compound-I-like species of P450. This species might abstract an electron from the substrate and release a proton to finish the O_2 transfer. Displacement by water completes the cycle. Of course, this mechanism is only speculative, with no intermediates beyond **7** having been observed for PDO or any other Rieske oxygenase.

Fe(II)–pterin-dependent hydroxylases

Almost 30 years ago, three enzymes that catalyse the aromatic ring hydroxylation of mandelate (reaction 5), benzoate (reaction 6) and anthranilate (reaction 7) were identified (reviewed in [23]). The activity of each was shown to be Fe(II) dependent and could be greatly stimulated by the addition of a reduced pterin cofactor. The method by which pterin enhances activity is not understood.

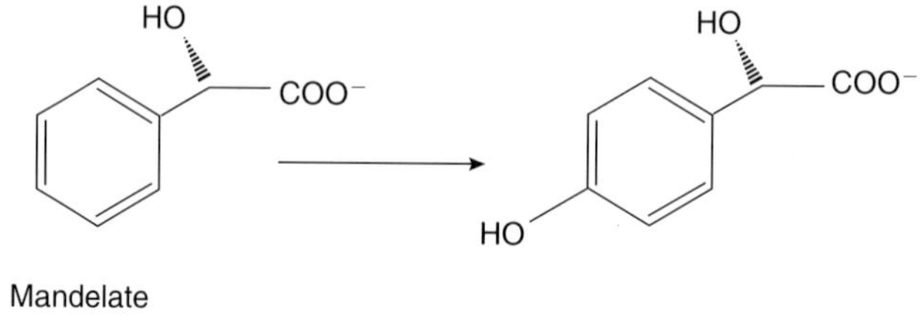

HO COO$^-$ → HO COO$^-$ HO (5)

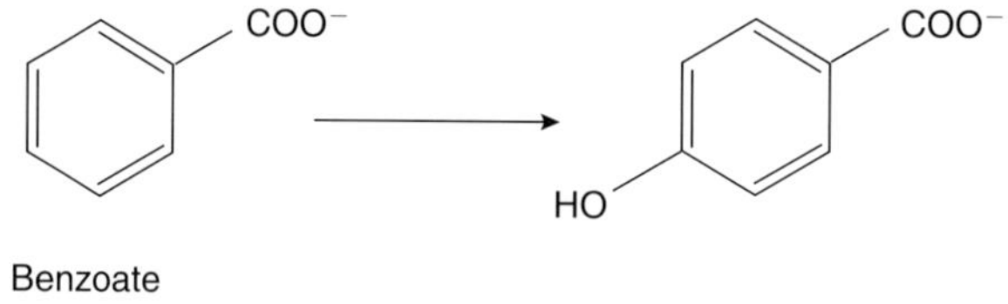

Mandelate

COO$^-$ → COO$^-$ HO (6)

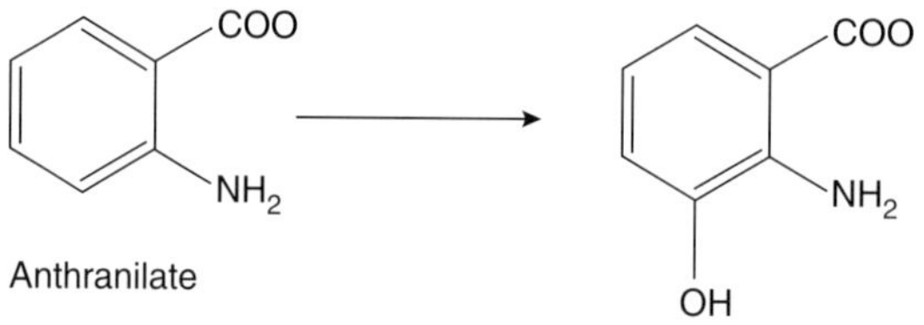

Benzoate

COO NH$_2$ → COO NH$_2$ OH (7)

Anthranilate

Relatively few reports have appeared in the literature since then. While not strictly involved in biodegradation of aromatic compounds, the aromatic amino acid hydroxylases, namely tyrosine, phenylalanine and tryptophan hydroxylases, also catalyse a single hydroxylation of an aromatic ring in an Fe(II)- and pterin-dependent fashion [23].

Structure of the iron centre

The first crystal structures of pterin-containing enzymes, tyrosine and phenylalanine hydroxylases, have recently been reported [15,16] (see Figure 5). The mononuclear Fe(II) site in each of these enzymes is ligated by an identical amino acid motif ($HX_4HX_{39}E$; which is similar to that of NDO). However, in the absence of amino acid sequences for the mandelate-, benzoate- or anthranilate-hydroxylating systems, it is impossible to know if a similar co-ordinating motif is present in these enzymes.

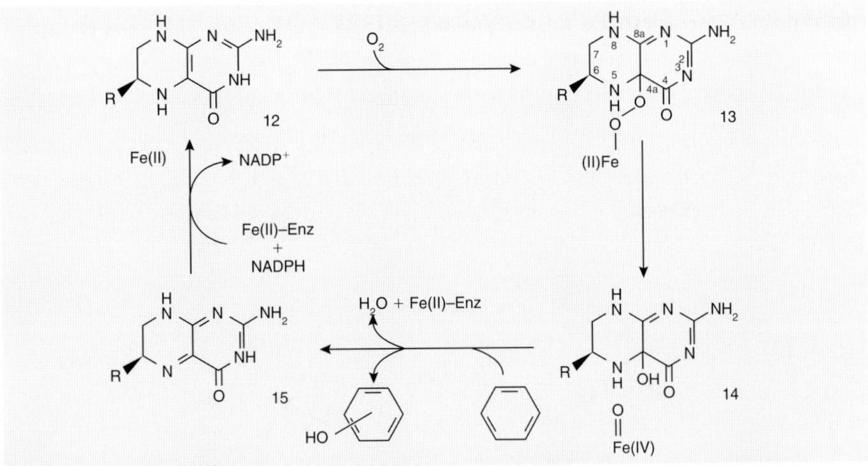

Figure 8. Proposed reaction cycle for the aromatic amino acid hydroxylases
Numbering of the ring is shown in 13. Enz, enzyme.

Mechanism

A mechanism for Fe(II)–pterin-supported hydroxylations may be proposed based on the current understanding of the aromatic amino acid hydroxylases (Figure 8). The active site of the resting form of the enzyme (**12**, see Figure 8) consists of the reduced tetrahydropterin, which serves as the source of electrons and probably participates in activating O_2, and the mononuclear iron in the Fe(II) oxidation state. Addition of O_2 is proposed to generate the Fe(II)–4a-peroxypterin intermediate (**13**). Heterolytic cleavage spawns a 4a-hydroxypterin (**14**) and a high-valency oxoferryl [Fe(IV)=O] species, which is proposed to be the active oxygenating agent, as proposed previously for the P450s and Rieske oxygenases. Upon oxygenation of the aromatic substrate, the 4a-hydroxypterin is released from the enzyme, and it loses water (often aided by a dehydratase) to form oxidized pterin (**15**). The pterin is re-reduced enzymically to tetrahydropterin, and then it binds again to the enzyme to participate in another round of catalysis. Although this reaction is classified as an intermolecular dioxygenation (reaction 3), the incorporation of one atom of oxygen into substrate while the other forms water (because the hydroxypterin dehydrates) makes it appear to be a mono-oxygenation (reaction 1). The mandelate, benzoate and anthranilate hydroxylases are likely to utilize similar mechanisms, but detailed studies of these enzymes have not been reported.

Di-iron-oxo mono-oxygenases

Non-haem, non-sulphur, di-iron-oxo clusters are involved in several mono-oxygenases; in addition, they participate in functions as diverse as reversible O_2 binding, iron regulation and production of reactive radicals in ribonucleotide reductases [24]. Crystal structures are available for several di-iron-containing proteins, including haemrythrin ([25]; see also Chapter 6 in this volume) and methane mono-oxygenases (MMOs) [26]. MMO is the most thoroughly studied di-iron mono-oxygenase [24]; other such oxygenases include toluene 2-, 3- and 4-mono-oxygenases, which catalyse *ortho-* (reaction 8), *meta-* (reaction 9) and *para*-hydroxylations (reaction 10), respectively (reviewed in [24]):

These di-iron-oxo-cluster-containing toluene mono-oxygenase systems constitute alternative pathways to those provided by the Rieske oxygenases for the biodegradation of toluene. Although methane is a small aliphatic compound, MMO has also been shown to exhibit activity towards toluene, catalysing several reactions analogous to the di-iron-oxo-cluster-containing toluene hydroxylases. The di-iron clusters of these mono-oxygenases must form potent oxidants to be able to carry out catalysis; for example, the conversion of methane to methanol catalysed by MMO is one of the most energetically demanding reactions in all of biology [24].

All of these systems have similar components, including redox-active proteins for transferring electrons from NADH to the oxygenase, and the oxygenase components have homologous sequences that are appropriate for forming di-iron-oxo centres. Three other related systems also utilizing di-iron-oxo centres are the alkane hydroxylase, certain phenol hydroxylases, and the stearoyl carrier protein Δ^9-desaturase. The latter forms *cis*-double bonds in fatty acids, utilizing NADPH and O_2, while releasing $2H_2O$. These are reviewed in [24].

Structure of the di-iron centre

In each of the di-iron clusters so far determined, the metal atoms are ligated by histidines and carboxylates in varying ratios [27]. Spectroscopic studies have confirmed that toluene 4-mono-oxygenase (reaction 10) contains a di-iron-oxo cluster [28]. Crystal structures are not available for toluene 2-, 3- or 4-mono-

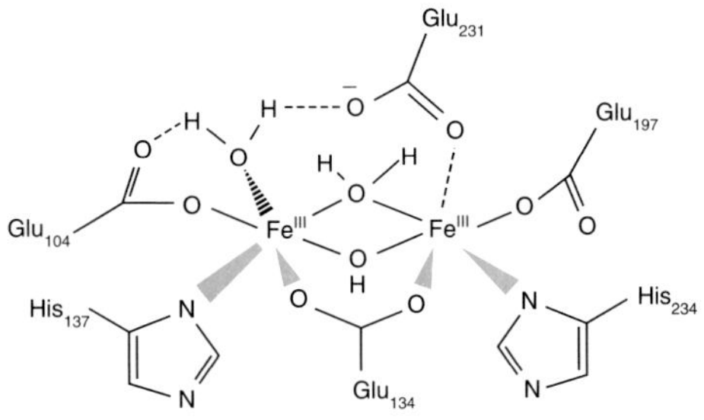

[-E104X_{29}E134X_2H137X_{59}E197X_{33}E231X_2H234-]-

Figure 9. Proposed structure for the di-iron-oxo site of toluene 4-mono-oxygenase based on the crystal structure of MMO [26] and the amino acid sequence of toluene 4-mono-oxygenase from *P. mendocina* [28]

oxygenases, but the crystal structure of soluble MMO has shown that the di-iron site ligands are contained in an amino acid motif ($EX_{29}EX_2HX_{61}EX_{33}EX_2H$) provided by the scaffold of a characteristic four-helix bundle [26]. Inspection of the toluene 4-mono-oxygenase sequence reveals a similar motif ($EX_{29}EX_2HX_{59}EX_{33}EX_2H$) [28]. A possible structural representation of the di-iron site of toluene 4-mono-oxygenase, based on the reported MMO structure and the amino acid sequence of the toluene 4-mono-oxygenase from *Pseudomonas mendocina*, is shown in Figure 9.

Mechanism

In recent years an incredible amount of mechanistic data has become available for MMO, whereas the toluene mono-oxygenases have only recently been identified [24]. It is expected that if toluene 4-mono-oxygenase and MMO have structurally similar di-iron sites, they may also have similar mechanisms. A possible mechanism for MMO-supported hydroxylation is presented in Figure 10 [29]. Initially, the diferric state (**16**, see Figure 10) is reduced by two electrons to form the diferrous state (**17**). This involves considerable change in the ligation structure, with the ligation at the irons transforming from 6-co-ordinate to 5-co-ordinate. Oxygen then binds to both iron atoms to generate the diferric peroxo intermediate P (for peroxo; **18**). Homolysis of the O–O bond results in the diamond core-structure intermediate Q (**19**), which may be thought of as consisting of two Fe(IV)=O units (**20**) [29]. This high-valency iron-oxo compound is thought to hydroxylate the hydrocarbon in a process analogous to that for cytochrome P450 (see Chapter 6 in this volume). The bridging carboxylate (E134 in Figure 9) is likely to be the keystone for maintaining the integrity of the di-iron centre during the ligation changes that are thought to occur during catalysis.

Di-iron-oxo enzymes also carry out hydroxylations of alkyl groups, such as side chains of aromatic compounds, as shown in reaction 11 [30,31], and desaturations of fatty acids (for example, stearoyl CoA to oleyl CoA) [32]. The

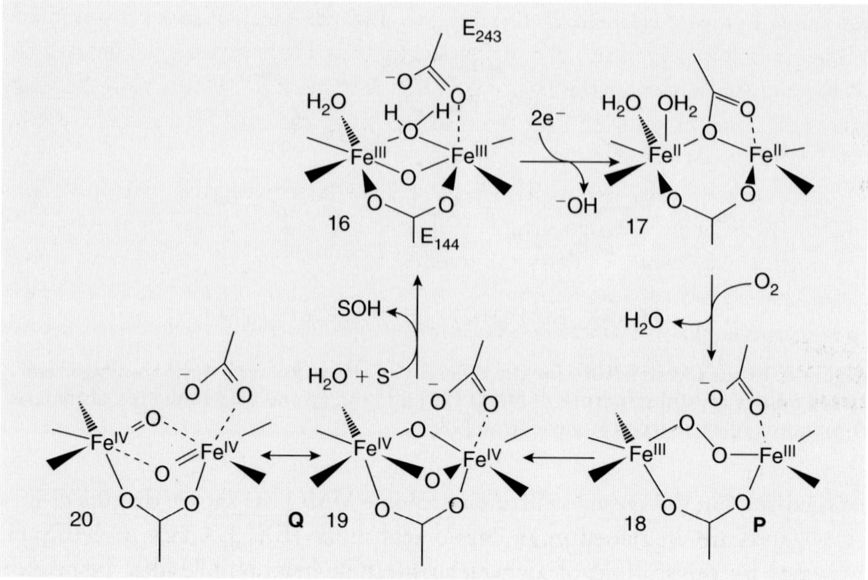

Figure 10. Proposed reaction cycle for di-iron-oxo mono-oxygenases [24,26,29]
S is substrate and SOH is the oxygenated product.

mechanisms are likely to be similar to those shown in Figure 10, with the Fe=O species prominent in the oxidations.

Conclusions

Oxygenases employing non-haem iron centres are prevalent in the bacterial degradation of aromatic compounds. Ferrous centres are frequently utilized to bind and activate O_2 for reaction with rather unreactive compounds. In all of the cases mentioned in this Chapter, it can be noticed that the ligands are principally histidines, carboxylates, amides and water. These ligands frequently form weak bonds with metals, and may thereby permit flexibility in the ligand environment necessary for the iron sites to undergo configurational and conformational changes during catalysis. Note that the mechanisms discussed above all invoke fairly substantial structural rearrangements around the iron during oxygenation. These types of rearrangement may be important in promoting these difficult dioxygenations and mono-oxygenations, as well as desaturation reactions at non-haem iron centres.

Perspectives

This review has focused on non-haem iron oxygenases involved in the biological degradation of aromatic compounds. The enzymes of these pathways exhibit rich diversity in substrate specificity, structure and mechanism. Whereas our understanding is constantly improving, much remains to be learned. Crystal structures have not been determined for many of the enzymes described in this review, and the mechanisms proposed rely strongly on the small number of structures available. Thus new structures will be invaluable in delineating more precise mechanisms. Spectroscopy has been and will continue to be important in detecting intermediates and elucidating structures of species in these reactions.

The potential of using these enzymes for bioremediation is tremendous; nearly all aromatic compounds can be metabolized by some micro-organism or another. Molecular biology is currently being employed to genetically engineer new breeds of 'super enzyme' with altered substrate specificity and more potent oxidizing potentials. Successful engineering of these enzymes will depend on the careful determination of the structural and functional roles of the proteins involved.

Summary

- *A wide variety of aromatic hydrocarbons can be degraded aerobically by micro-organisms. A large fraction of the metabolic pathways are initiated by oxygenases containing Fe(II) at the active sites, which participates in the oxygenation and activation of the hydrocarbons.*
- *Mono-oxygenations and dioxygenations are found in these pathways. Some of these enzymes can catalyse either or both reactions, depending on the nature of the substrate.*
- *Two general themes are found: mononuclear Fe(II) centres that must be reduced by one electron at a time, or di-iron centres that can be reduced by two electrons. The electrons from NAD(P)H can be delivered by either an electron-transfer chain consisting of a flavin and one or more [2Fe–2S] centres, or a pterin.*
- *Proposed mechanisms generally involve higher oxidation states of the iron (Fe=O), analogous to those for P450, and peroxidase systems. These strong oxidants are necessary to oxidize aromatic and aliphatic compounds.*
- *Mechanisms currently considered viable for these reactions require significant changes in ligation during catalysis. The structures of the non-haem iron centres may be particularly well-suited for such transformations.*

We are grateful for a grant from the National Institutes of Health, no. GM20877.

References

1. Harwood, C.S. & Gibson, J. (1997) Shedding light on anaerobic benzene ring degradation: a process unique to prokaryotes? *J. Bacteriol.* **179**, 301–309
2. Que, L. & Ho, R.Y.Y. (1996) Dioxygen activation by enzymes with mononuclear non-heme iron active sites. *Chem. Rev.* **96**, 2607–2624

3. Butler, C.S. & Mason, J.R. (1997) Structure-function analysis of the bacterial aromatic ring-hydroxylating dioxygenases. *Adv. Microb. Physiol.* **38**, 277–305
4. Batie, C.J. & Ballou, D.P. (1990) Phthalate dioxygenase. *Methods Enzymol.* **188**, 61–70
5. Correll, C.C., Batie, C.J., Ballou, D.P. & Ludwig, M.L. (1992) Phthalate dioxygenase reductase: a modular structure for electron transfer from pyridine nucleotide to [2Fe–2S]. *Science* **258**, 1604–1610
6. Gurbiel, R.J., Doan, P.E., Gassner, G.T. et al. (1996) Active site structure of Rieske type proteins: electron nuclear double resonance studies of isotopically labeled phthalate dioxygenase from *Pseudomonas cepacia* and Rieske protein from *Rhodobacter capsulatus* and molecular modeling studies of a Rieske center. *Biochemistry* **35**, 7834–7845
7. Kauppi, B., Lee, K., Carredano, E., Parales, R.F., Gibson, D.T., Eklund, H. & Ramaswamy, S. (1998) Structure of an aromatic-ring-hydroxylating dioxygenase-naphthalene dioxygenase. *Structure* **6**, 571–586
8. Keyser, P.K., Pujar, B.G., Eaton, R.W. & Ribbons, D.W. (1976) Biodegradation of the phthalates and their esters by bacteria. *Environ. Health Perspect.* **18**, 159–166
9. Tsang, H.-T., Batie, C.J., Ballou, D.P. & Penner-Hahn, J.E. (1996) Structural characterization of the mononuclear iron site in *Pseudomonas cepacia* DB01 phthalate dioxygenase using X-ray absorption spectroscopy. *J. Biol. Inorg. Chem.* **1**, 24–33
10. Han, S., Eltis, L.D., Timmis, K.N., Muchmore, S.W. & Bolin, J.T. (1995) Crystal structure of the biphenyl-cleaving extradiol dioxygenase from a PCB-degrading pseudomonad. *Science* **270**, 976–980
11. Gillmar, S.A., Villaseñor, A., Fletterick, R. et al. (1997) The structure of mammalian 15-lipoxygenase reveals similarity to the lipases and the determination of substrate specificity. *Nat. Struct. Biol.* **4**, 1003–1009
12. Boyington, J.C., Gaffney, B.J. & Arnzel, M. (1993) The three-dimensional structure of arachidonic acid 15-lipoxygenase. *Science* **260**, 1482–1486
13. Minor, W., Steczko, J., Stec, B. et al. (1996) Crystal structure of soybean lipoxygenase L-1 at 1.4 Å resolution. *Biochemistry* **35**, 10687–10701
14. Roach, P.L., Clifton, I.J., Hensgens, C.M. et al. (1997) Structure of isopenicillin N synthase with substrate and the mechanism of penicillin formation. *Nature (London)* **387**, 827–830
15. Goodwill, K.E., Sabatier, C., Marks, C. et al. (1997) Crystal structure of tyrosine hydroxylase at 2.3 Å and its implications for inherited neurodegenerative diseases. *Nat. Struct. Biol.* **4**, 578–585
16. Erlandsen, H., Fusetti, F., Martinez, A. et al. (1997) Crystal structure of the catalytic domain of human phenylalanine hydroxylase reveals the structural basis for phenylketonuria. *Nat. Struct. Biol.* **4**, 995–1000
17. Neidle, E.L., Hartnett, C., Ornston, L.N. et al. (1991) Nucleotide sequences of the *Acinetobacter calcoaceticus benABC* genes for benzoate 1,2-dioxygenase reveal evolutionary relationships among multicomponent oxygenases. *J. Bacteriol.* **173**, 5385–5395
18. Jiang, H., Parales, R.E., Lynch, N.A. & Gibson, D.T. (1996) Site-directed mutagenesis of conserved amino acids in the alpha-subunit of toluene dioxygenase: potential mononuclear non-heme iron co-ordination sites. *J. Bacteriol.* **178**, 3133–3139
19. Coulter, E.D., Moon, N., Batie, C.J. et al. (1999) Probing the mononuclear Fe(II) site of phthalate dioxygenase by electron paramagnetic resonance spectroscopy: direct evidence for ligation of two histidines in the Cu(II)-reconstituted protein. *Biochemistry*, in the press
20. Gibson, D.T., Resnick, S.M., Lee. K. et al. (1995) Desaturation, dioxygenation, and monooxygenation reactions catalysed by naphthalene dioxygenases *Pseudomonas* sp. strain 9816-4. *J. Bacteriol.* **177**, 2615–2621
21. Bernhardt, F.-H., Bill, E., Trautwein, A.X. & Twilfer, H. (1988) 4-Methoxybenzoate monooxygenase from *Pseudomonas putida*: isolation, biochemical properties, substrate specificity, and reaction mechanisms of the enzyme components. *Methods Enzymol.* **161**, 281–294

22. Gassner, G.T., Ballou, D.P., Landrum, G.A. & Whittaker, J.W. (1993) Magnetic circular dichroism studies of the mononuclear ferrous active site of phthalate dioxygenase from *Pseudomonas cepacia* show a change of ligation state on substrate binding. *Biochemistry* **32**, 4820–4825
23 Kappock, T.J. & Caradonna, J.P. (1996) Pterin-dependent amino acid hydroxylases. *Chem. Rev.* **96**, 2659–2756
24. Wallar, B.J. & Lipscomb, J.D. (1996) Dioxygen activation by enzymes containing binuclear non-heme iron clusters. *Chem. Rev.* **96**, 2625–2657
25. Stenkemp, R.E., Sieker, L.C., Jensen, L.H. & McQueen, J.E. (1978) Structure of methemerythrin at 2.8 Å resolution: computer graphic fit of an averaged electron density map. *Biochemistry* **17**, 2499–2504
26. Rosenzweig, A.C., Frederick, C.A., Lippard, S.J. & Nordlund, P. (1993) Crystal structure of a bacterial non-haem iron hydroxylase that catalyses the biological oxidation of methane. *Nature (London)* **366**, 537–543
27. Kurtz, D.M. (1997) Structural similarity and functional diversity in diiron-oxo proteins. *J. Biol. Inorg. Chem.* **2**, 159–167
28. Pikus, J.D., Studts, J.M., Achin, C. et al. (1996) Recombinant toluene-4-monooxygenase: catalytic and Mössbauer studies of the purified diiron and Rieske components of a four protein complex. *Biochemistry* **35**, 9106–9119
29. Shu, L., Nesheim, J.C., Kaufmann, K. et al. (1997) An $Fe_2^{IV}O_2$ diamond core structure for the key intermediate Q of methane monooxygenase. *Science* **275**, 515–518
30. Suzuki, M., Hayakawa, T., Shaw, J.P. et al. (1991) Primary structure of xylene monooxygenase: similarities to and differences from the alkane hydroxylation system. *J. Bacteriol.* **173**, 1690–1695
31. Shanklin, J., Achim, C., Schmidt, H. et al. (1997) Mössbauer studies of alkane Ω-hydroxylase: evidence for a diiron cluster in an integral-membrane enzyme. *Proc. Natl. Acad. Sci. U.S.A.* **94**, 2981–2986
32. Shanklin, J., Whittle, E. & Fox, B.G. (1994) Eight histidine residues are catalytically essential in a membrane-associated iron enzyme, stearoyl-CoA desaturase, and are conserved in alkane hydroxylase and xylene monooxygenase. *Biochemistry* **33**, 12787–12794

4

Haem iron-containing peroxidases

Issa S. Isaac* and John H. Dawson*†[1]

**Department of Chemistry and Biochemistry, University of South Carolina, Columbia, SC 29208, U.S.A., and †School of Medicine, University of South Carolina, Columbia, SC 29208, U.S.A.*

Introduction

Haem-containing peroxidase enzymes are widely distributed throughout the plant and animal kingdoms and also have been isolated from bacteria, mould and micro-organisms [1]. Peroxidases serve the role of antioxidants, protecting cells, tissues and organs against the toxic effects of peroxides produced *in vivo* by oxidase activity [1]. They catalyse the oxidation of a variety of organic and inorganic compounds by hydrogen peroxide, organic hydroperoxides, peracids or inorganic oxides, such as periodate or chlorite. The general peroxide-dependent reaction catalysed by a peroxidase is shown in eqn. (1):

$$ROOH + 2AH_2 \xrightarrow{\text{Peroxidase}} 2AH^{\bullet} + ROH + H_2O \qquad (1)$$

So far, 13 haem-containing peroxidase crystal structures have been reported, including *Arthromyces ramosus* peroxidase [2], ascorbate peroxidase [2], peanut peroxidase [2], *Coprinus cinereus* peroxidase [3], cytochrome *c* peroxidase (CCP) [3], lignin peroxidase [3], myeloperoxidase [3], prostaglandin H_2 synthase [3], *Caldariomyces fumago* chloroperoxidase (CPO) [4], manganese peroxidase [5], di-haem CCP from *Pseudomonas aeruginosa* [6] and, most recently, horseradish peroxidase (HRP) isozyme C [2] and barley-grain peroxidase [5]. All of these peroxidases contain the haem (iron protoporphyrin IX) prosthetic group with a histidine proximal ligand, with

[1] *To whom correspondence should be addressed.*

the exception of the cysteinate-ligated CPO. The overall folding patterns of the histidine-ligated peroxidases are quite similar despite relatively little similarity of their primary sequences [7].

Peroxidase reaction sequence

HRP has been used as the paradigm for the biochemical study of peroxidases. It is the most extensively studied member of the plant peroxidase superfamily [1]. The peroxidase mechanism to be discussed below has been derived primarily from studies of HRP.

The peroxidase mechanism comprises four steps, as shown in Figure 1. The native ferric state (structure **1**, see Figure 1) reacts with hydrogen peroxide on the distal side of the haem to generate a transient hydroperoxide adduct, compound 0 (**2**), which breaks down rapidly (so fast that it cannot normally be observed) to give an oxoferryl (sometimes called oxyferryl) porphyrin π-cation radical, compound I (**3**). Peroxidases can also utilize organic hydroperoxides, peracids and other oxygen-atom donors in place of hydrogen peroxide to generate compound I. Next, compound I is reduced by an electron-rich substrate (AH_2) to generate compound II (**4**) and a cation radical of the sub-

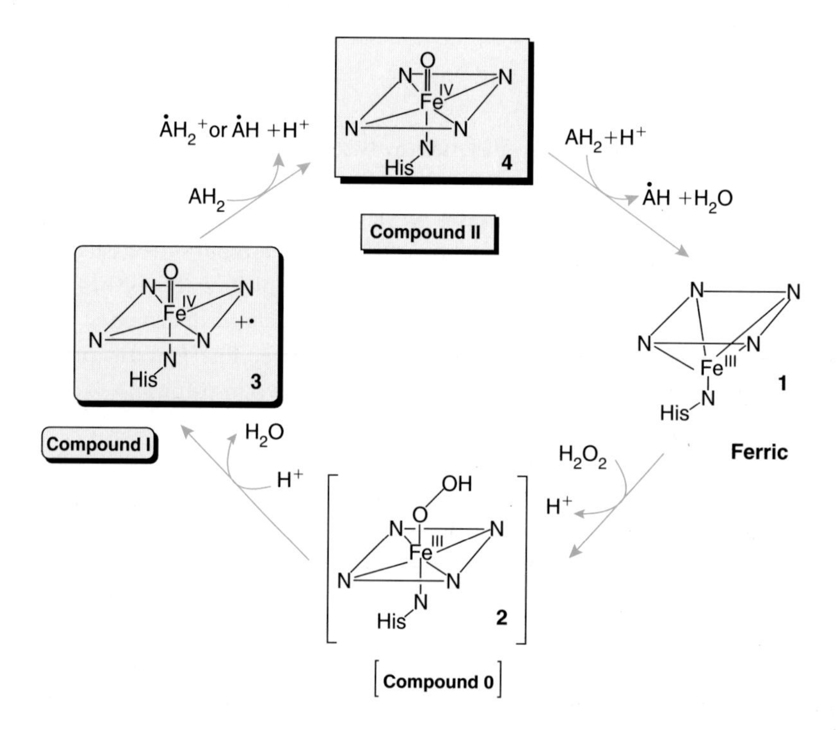

Figure 1. Peroxidase cycle
The porphyrin ring of the haem is represented by a parallelogram with the four pyrrole nitrogens at the corners.

strate, which, depending on its acidity, will release H^+. Alternatively, AH_2 can contribute a hydrogen atom ($H^{\cdot}$) to produce $AH^{\cdot}$ and Fe(IV)-OH;the latter can lose a proton to become Fe(IV)=O. This will be discussed later. A second AH_2 can then react with compound II to regenerate the ferric resting state (**1**). In this reaction, the substrate contributes a $H^{\cdot}$ (or an electron plus a proton), and the resulting Fe(III)-OH is protonated and releases H_2O. The two radical-product molecules disproportionate non-enzymically to form a two-electron oxidized-product molecule (A) and a regenerated substrate molecule (AH_2), see eqn. (2). The rate of substrate oxidation by compound I of HRP is usually 10–100-fold faster than that by compound II [1]. This may be due to reorganizational energy differences between the two processes. The reduction of compound I to II merely requires the delivery of an electron to the porphyrin π-cation radical. However, the reduction of compound II to the ferric state requires the delivery of an electron and two protons (or a hydrogen atom plus a proton), so that H_2O can be released. In this process, the haem undergoes a reorganization to 5-co-ordinate iron.

$$AH^{\cdot} + AH^{\cdot} \longrightarrow AH_2 + A \quad (2)$$

The peroxidase catalytic reaction cycle, illustrated with structures of the haem prosthetic group in Figure 1, involves two high-valency oxoferryl intermediates. Compound I (**3**) is two oxidation equivalents above the native ferric state of the haem, and compound II (**4**) is one oxidation equivalent above native ferric haem (**1**). The ferric–hydroperoxide complex (**2**), designated compound 0, is thought to be formed after mixing HRP and hydrogen peroxide *en route* to the generation of compound I.

CPO purified from the mould *C. fumago* stands out among the peroxidases due to its versatile reactivities. Along with the typical peroxidase activity, CPO has a unique ability among peroxidases to catalyse the chloride ion-dependent chlorination of certain organic substrates [8]. CPO utilizes intermediates similar to those observed with HRP to catalyse three distinctly different peroxide- or peracid-dependent reactions (Figure 2): (i) halogenation of organic compounds with Cl^-, Br^- or I^- as the halogen source (**5**→**6**→**7**→**9**→**5**; see Figure 2); (ii) a typical peroxidase activity (**5**→**6**→**7**→**8**→**5**); and (iii) a catalase (disproportionation of hydrogen peroxide) activity (**5**→**6**→**7**→**5**). The halogenation reaction is thought to proceed through compound X (**9**), a ferric hypochlorite species.

Whereas the proximal ligand of most other peroxidases is the nitrogen of a histidine residue, CPO is ligated by the anionic sulphur of a deprotonated cysteine (cysteinate) residue [8]. This unique property of CPO has qualified it as a particularly useful protein model for reactions of cytochrome P450, another cysteinate-ligated haem-containing enzyme (see Chapter 5 in this volume). The enzymes of the cytochrome P450 superfamily consist of mono-oxygenases that activate molecular oxygen to catalyse an extensive variety of oxygen-transfer reactions ([8] and Chapter 5 in this volume).

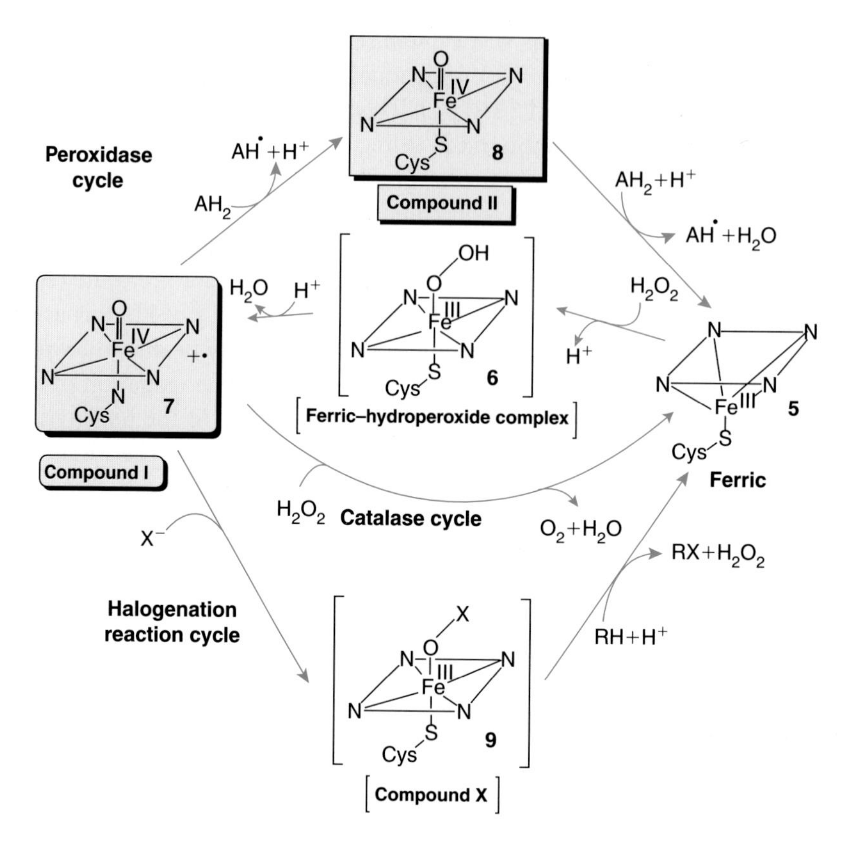

Figure 2. Reaction cycle of CPO

Compound I

Physical properties

This compound has a characteristic green colour with an oxidation state two equivalents above the ferric enzyme [1,3]. Establishing the distribution of electrons around the haem, the iron and the protein in compound I has been a challenge. Three resonance forms could contribute to its structure (structures **3**, **3a** and **3b**, see Figure 3). Figure 3 also displays an alternative structure, **3c**, for the compound I of CCP (originally called compound ES, CCP-ES) that will be discussed below.

HRP

Various lines of evidence have led to the conclusion that structure **3**, a green oxoferryl porphyrin π-cation radical, is the predominant structure of HRP compound I (HRP-I) [9,10]. The UV-visible absorption spectrum of HRP-I features a Soret band at 400 nm that is only about half the intensity of the native HRP Soret peak at 402 nm (Figure 4a). The loss of absorbance intensity is attributed to the loss of aromaticity in the porphyrin on forming the

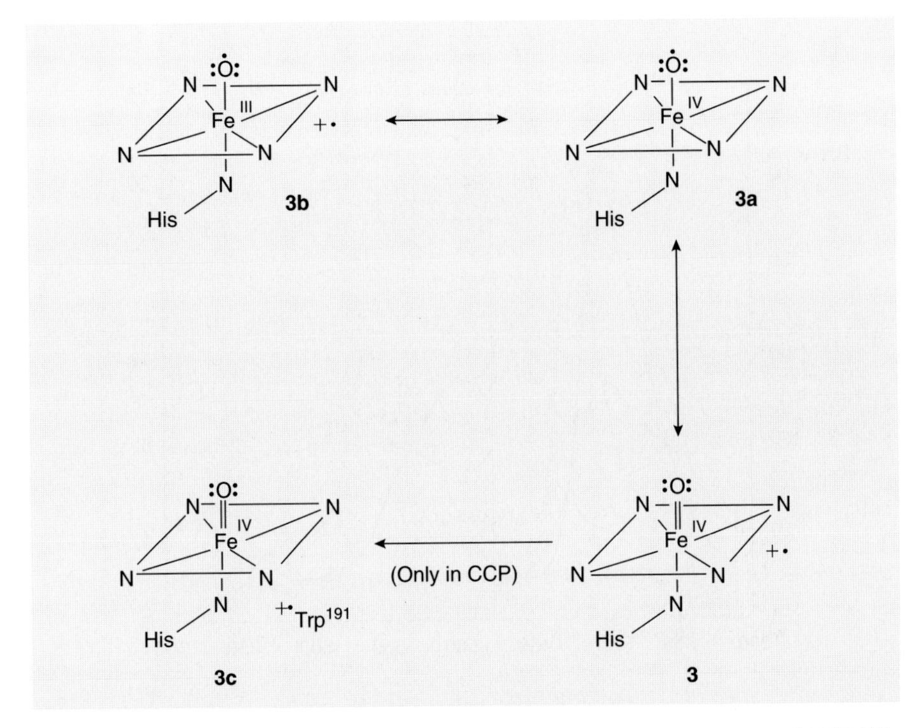

Figure 3. Resonance structures of HRP-I (3, 3a and 3b) and the structure of CCP-ES (3c)
The formal oxidation state of the haem is V, and the total charge of the species is +1.

π-cation radical. In Figure 4, the UV-visible absorption spectra of other HRP intermediates (i.e. HRP-II and HRP-III) and CCP-ES are also displayed. In the visible region, peaks present at 498 and 640 nm in the spectrum of native ferric HRP disappear upon formation of compound I, leaving a broad series of weakly absorbing peaks. The lack of distinct bands in the visible region is generally cited as evidence for the presence of a porphyrin radical [8].

Magnetic susceptibility measurements have shown that HRP-I is paramagnetic with three unpaired electrons ($S=3/2$), suggesting a formal iron oxidation state of V for HRP-I [1]. However, Mössbauer spectroscopy [11] showed that the oxidation state of the haem iron in HRP-I is Fe(IV), implying that the fifth oxidation equivalent is not on the iron. Extended X-ray absorption fine structure spectroscopy (EXAFS) of the haem iron site in HRP-I demonstrated that the Fe–O bond length is 1.64 Å [10,12]. This short bond for HRP-I is consistent with an oxoferryl complex [Fe(IV)=O]. Moreover, electron-nuclear double resonance spectroscopy (ENDOR) confirmed that HRP-I contains an oxo ligand [13,14].

Work on both protein and model compounds supports the conclusion that one of the oxidation equivalents resides on the haem as a π-cation radical. Resonance Raman [15] and NMR [16] data, as well as iterative extended Hückel calculations [17], have all substantiated that a π-cation radical actually resides

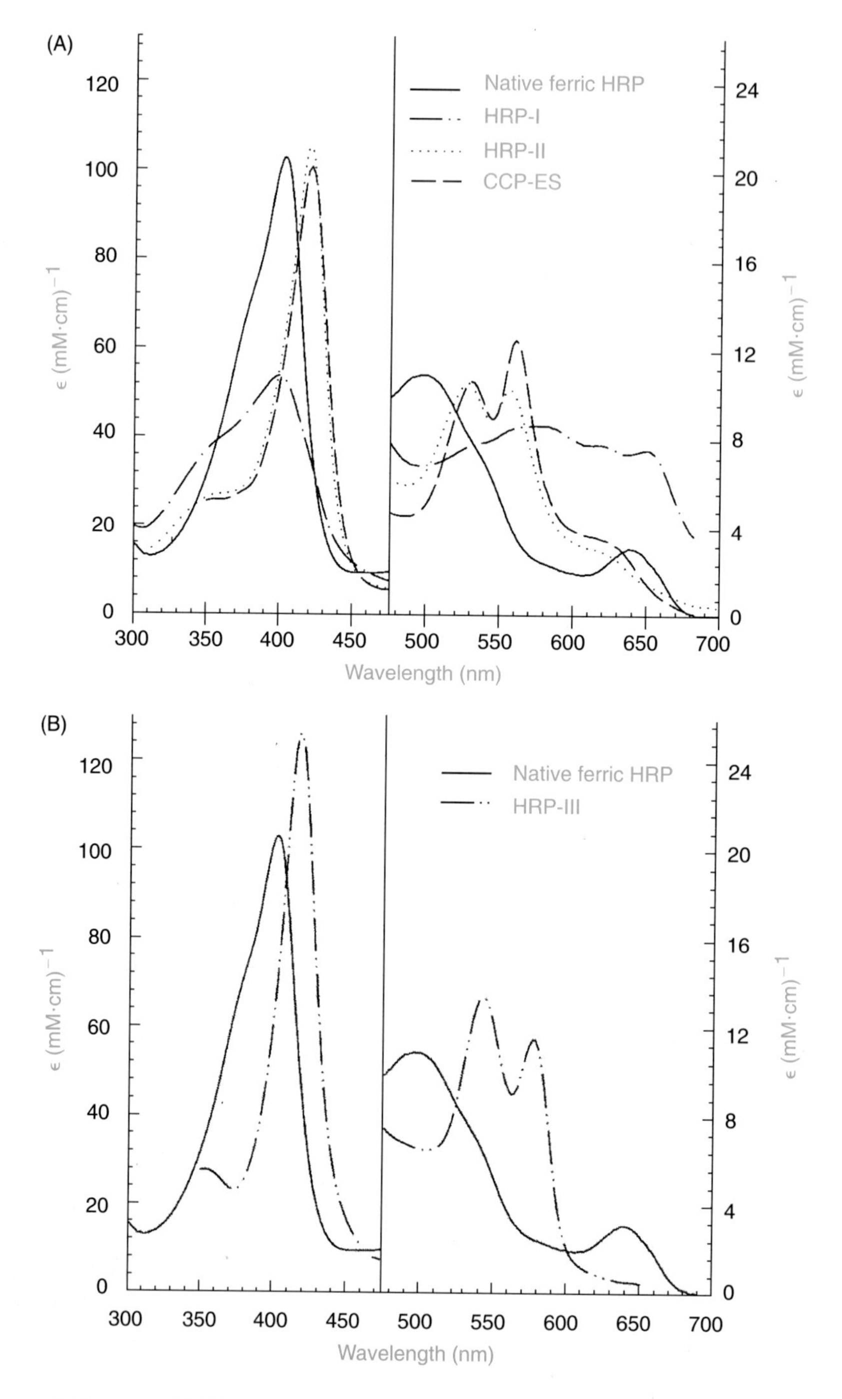

Figure 4. Spectra of HRP species in various states
(A) UV-visible absorption spectra of HRP species and CCP-ES at neutral pH, and (B) UV-visible absorption spectra of HRP species at neutral pH. Spectra were recorded at 4°C at pH7.0 in 100 mM potassium phosphate buffer.

on the haem. The Mössbauer and EPR data for HRP-I [18] show that the $S=1$ spin of the Fe(IV)=O group is weakly coupled to a spin $S'=1/2$ radical on the haem. EPR studies of Zn-substituted HRP further demonstrate that the porphyrin can be oxidized to a π-cation radical with hexachloro-iridate [9]. In contrast to the iron, the Zn cannot be oxidized, so that the oxidation equivalent has to reside on the ring. This Zn-porphyrin radical has a visible spectrum, which is very similar to normal compound I that contains iron, including the same lower-intensity optical Soret band; thus the loss of aromaticity of the porphyrin ring upon oxidation is associated with the lower-intensity Soret band.

CCP

In most peroxidases, the haem group in compound I is in the +5 oxidation state (ferryl iron coupled to the porphyrin radical). However, with CCP, rapid intramolecular electron transfer from a nearby amino acid residue (Trp-191) to the nascent porphyrin π-cation radical generates a stable compound, ES (**3c**, Figure 3, and see Figure 7 later). This complex, which is two oxidizing equivalents above the ferric state, consists of a protein-based tryptophan cation radical and an oxoferryl haem centre equivalent to HRP-II [19]. As expected, the UV-visible absorption (Figure 4a) and magnetic CD spectra of CCP-ES are different from those of HRP-I, but are similar to those of HRP-II [20]. EXAFS measurements [21] show that the Fe–O bond length in CCP-ES is 1.67 ± 0.04 Å, which is characteristic of the oxoferryl complex, Fe(IV)=O, in HRP-I [10,12]. In addition, resonance Raman spectroscopy verified the presence of an Fe(IV)=O bond in CCP-ES [22]. The crystal structure of this intermediate confirmed the presence of an oxoferryl complex [Fe(IV)=O] [19].

Identification of the site of the amino acid radical in compound ES was more elusive. Unlike HRP-I, which is green, CCP-ES is red and exhibits an EPR signal typical of a free radical, suggesting that a free radical resides on an amino acid [23]. Initially, several amino acids were considered to harbour the free radical, although further investigations placed the free radical on Trp-191 in the active site [3]. EPR of CCP mutants, with Trp-51 and Trp-191 separately replaced by Phe, showed that Trp-191, rather than Trp-51, was the site of the tryptophan radical [3]. ENDOR studies on mutants containing perdeuterated Trp residues showed that the second oxidizing equivalent resided on Trp-191 in the wild-type CCP-ES [24]. Furthermore, when the Trp-191→Phe mutant was mixed with hydrogen peroxide in a stopped-flow spectrophotometer, a porphyrin π-cation radical was observed with a half life of ≈14 ms [3]. Recent ENDOR data have confirmed that CCP-ES consists of an oxoferryl haem coupled by weak exchange to the Trp-191 π-cation radical [25].

CPO

Compound I of CPO (CPO-I) exhibits a Soret band at 368 nm and a distinguishing peak at 690 nm. Like other peroxidases, CPO can form compound I upon reaction with peracids and organic hydroperoxides [26].

The weak intensity of the Soret band and the lack of distinct bands in the visible region of the spectrum of CPO-I indicate the formation of a porphyrin π-cation radical [8]. The EPR and Mössbauer spectroscopy of CPO-I are indicative of a porphyrin π-cation radical with a stronger coupling to the haem ferryl iron than is the case with HRP-I [8]. The properties are consistent with the model of an exchange-coupled spin $S=1$ Fe(IV) iron and a spin $S'=1/2$ porphyrin radical [8], like that of HRP.

The nature of the amino acid residues and the environment that is present in the pocket containing the haem, on both the proximal and distal sides, is likely to be of great influence on the structure and reactivity of intermediates, including compound I. The proximal ligand appears to be less important than other factors. In CCP, for example, the proximal histidine ligand was mutated to glutamine with no decrease in the rate of reaction with hydrogen peroxide; spectroscopically, both the ferric and compound-I states of this mutant closely resemble those of the wild-type enzyme, presumably due to ligation of the oxygen of the glutamine side chain to haem iron. However, the stability of compound I was drastically reduced [27].

Urano et al. have suggested that the oxygen-radical resonance form of compound I (**3a**, Figure 3) is more likely to be able to abstract a hydrogen atom from the substrate than is a porphyrin oxoferryl π-cation radical (**3**, Figure 3) [28]. In fact, CPO-I and HRP-I are even able to abstract a hydrogen atom from organic oxidants, such as alkyl hydroperoxides and peracids, as will be discussed later in more detail. Consequently, it is reasonable to ask which of the resonance structures of compound I displayed in Figure 3 are more energetically favoured and what are the determining factors?

Champion [29] proposed that the presence of a cysteinyl sulphur on the proximal side, as found in cytochrome P450 and CPO-I, favours the oxygen-radical-type species (**7a**, Figure 5) over the oxoferryl π-cation-radical form (**7**, Figure 5), and Urano et al. [28] used synthetic models to obtain experimental evidence for this proposal. Loew et al. [30] presented calculations for cytochrome P450 indicating that an oxo-radical species would be within 0.7 kcal/mol of the ground state. They speculated that this oxo-radical species is apparently stabilized as the iron–oxygen bond is lengthened.

From his study with cytochrome P450, Champion argued that the sulphur→iron π-electron donation appears to have mechanistic consequences for the O–O cleavage [29]. In thiolate-ligated enzymes, compound I has a neutral overall charge due to the negative charge on the proximal sulphur, thus favouring the oxo-radical configuration over the higher energy π-cation radical on the haem (Figure 5, top). However, in histidine-ligated enzymes, the presence of nitrogen with a neutral charge results in an overall charge of +1 for compound I (Figure 5, bottom). The formation of a π-cation radical is required to diminish the unfavourable effect of this charge by delocalizing it over the extensive π-bonding system of the haem. Thus, the π-cation radical form (**3**, Figures 3 and 5) is favoured over an oxo-radical type (**3a**, Figures 3

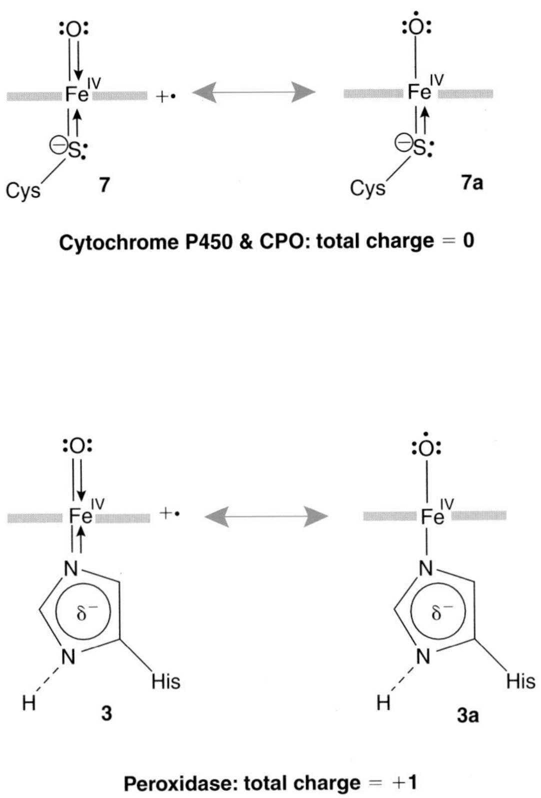

Cytochrome P450 & CPO: total charge = 0

Peroxidase: total charge = +1

Figure 5. Resonance structures for compound I showing the influences of the proximal ligand

and 5). Moreover, the absence of the sulphur allows the Fe(dπ) orbitals to couple more effectively with the oxygen, contributing to a shorter Fe–O bond [29]. The π-cation radical is more likely to be reduced by electron transfer, whereas the oxo-radical configuration is more likely to abstract a hydrogen atom. Nevertheless, Mössbauer and EPR data suggest that CPO, which has a proximal cysteine ligand, favours the π-cation radical configuration [8]. Champion suggests that in contrast to cytochrome P450, where the distal side contains primarily non-polar residues, CPO contains polar residues to help stabilize the π-cation radical.

Reactivity

Compound I is an intermediate that has been proposed to be directly involved in the enzymic mechanism of many haem-containing enzymes. In cytochrome P450, for example, it is thought to be the reactive intermediate responsible for abstracting a hydrogen atom from the organic substrate [8]. CPO-I, formed by reaction of the ferric enzyme with an alkyl hydroperoxide, has been shown by

EPR spectroscopy to abstract a hydrogen atom from a second alkyl hydroperoxide (ROOH) molecule to form compound II and a peroxyl radical (ROO$^{\bullet}$) [31]. Spin traps were used to detect the peroxyl radicals and the molecules formed from the breakdown of those radicals. Compound II then abstracts a hydrogen atom from a third molecule of alkyl hydroperoxide to generate the native ferric enzyme. Figure 6 illustrates the reaction sequence proposed by Chamulitrat et al. [31]. Similar conclusions for the reaction of CPO with excess peracid were obtained directly using stopped-flow spectrophotometry [26].

A spectroscopic and kinetics study produced results with HRP [32] that are also consistent with the reaction sequence described by Chamulitrat et al. for CPO [31]. These experiments showed that a peracid (metachloroperbenzoic acid) will react with HRP-I to generate compound II and an organic radical. The latter either diffuses away or, approximately 30% of the time, inactivates the enzyme by alkylating the haem.

The results obtained for the reactions of CPO and HRP with alkyl hydroperoxides or peracids demonstrate both the oxidizing potency of

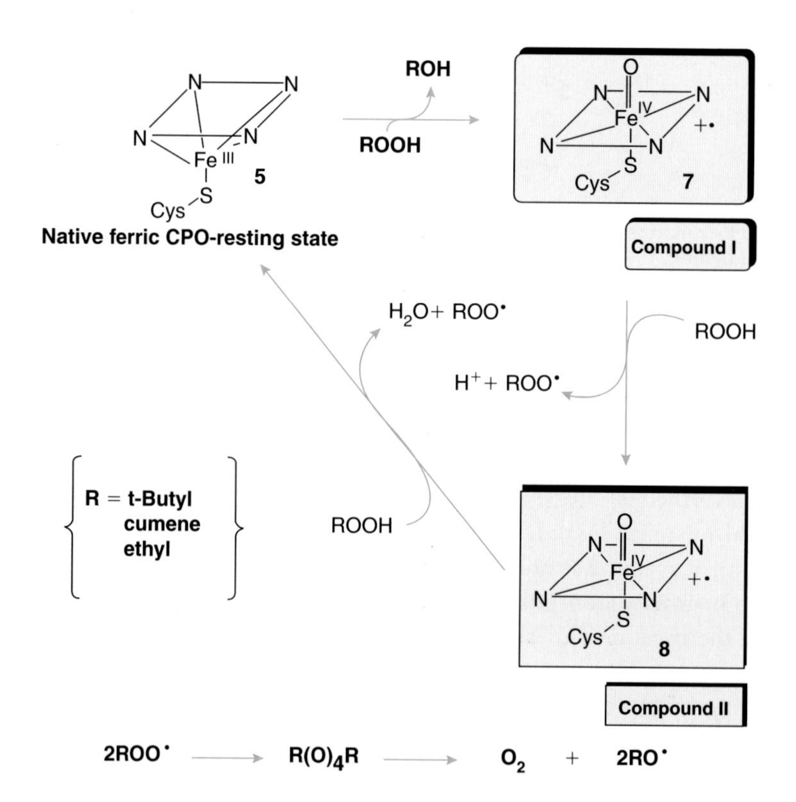

Figure 6. Scheme for the reaction of CPO with alkyl hydroperoxide

compound I and its ability to abstract a hydrogen atom (or an electron and a proton), not only from organic substrates, but also from either alkyl hydroperoxides or peracids [26]. In keeping with this, the oxidation potentials for compound I/compound II and for compound II/Fe(III) of HRP have each been estimated to be approximately +1 V (versus the normal hydrogen electrode) [33].

Mechanism of formation

HRP-I is readily formed when either hydrogen peroxide or any of a number of alternative oxygen-atom donors, such as organic hydroperoxides and peracids, is mixed with ferric HRP. The proposed pathway for the formation of HRP-I from an alkyl hydroperoxide is illustrated in Figure 7. Initially, the distal histidine residue acts as a general base to form a ferric alkyl hydroperoxide, **2a**. Evidence for formation of such a species will be discussed below. The positively charged distal arginine is thought to stabilize the developing negative charge on the β-oxygen during heterolytic O–O bond scission (**2b**). The transiently protonated histidine acts as a general acid to protonate the leaving alkoxide and generate compound I (**3**), and ROH is released from the active site. As described above, the distal histidine and

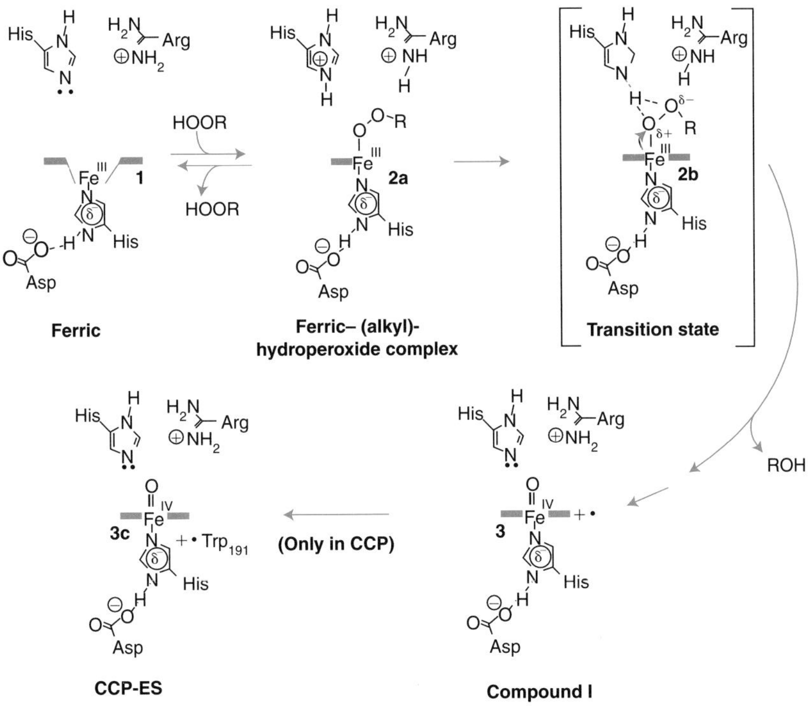

Figure 7. Formation of HRP-I

arginine residues play significant roles in the formation of HRP-I. The exact role of the proximal histidine ligand is still being examined [27].

Compound II

Physical properties

HRP-II has a characteristic red colour with a Soret absorption band at approximately 420 nm (Figure 4a). Magnetic-susceptibility measurements of compound II are consistent with the presence of two unpaired electrons [1] and Mössbauer spectra confirm that HRP-II indeed contains Fe(IV) [1]. The initial EXAFS study of HRP-II by Penner-Hahn et al. reported that the Fe–O bond was very short, approximately 1.6 Å [10,12]. EXAFS results reported by Chance et al. suggested that the length of the Fe–O bond was pH dependent, short at alkaline pH, and ≈1.93 Å at neutral pH [34]. Possible structures for HRP-II at neutral and alkaline pHs based on these EXAFS results are shown in Figure 8 (**4a** and **4**, respectively). Sitter et al. also observed pH-dependent properties in the resonance Raman spectra of HRP-II [35]. However, the magnitude of the shift in the Fe–O vibrational frequency as a function of pH was too small to indicate complete conversion from an Fe=O double-bonded unit to an Fe–OH single-bonded ligand. Thus it was proposed that, above pH 8.7, the oxygen of the oxoferryl species was hydrogen-bonded to an imidazolium hydrogen atom from the distal histidine residue in the active site. A non-hydrogen-bonded oxoferryl structure, Fe(IV)=O, was proposed as the predominant structure below pH 8.7 [35]. Further investigations will be necessary to better establish the structure of HRP-II as a function of pH.

Mechanism of formation

Compound I is reduced to compound II upon interaction with a substrate in the active site. One suggestion is that a hydrogen atom is transferred from the substrate to the oxoferryl oxygen/porphyrin π-cation-radical unit. The electron from the hydrogen atom would then reduce the porphyrin π-cation radical and the remaining proton would protonate the distal histidine [3,36],

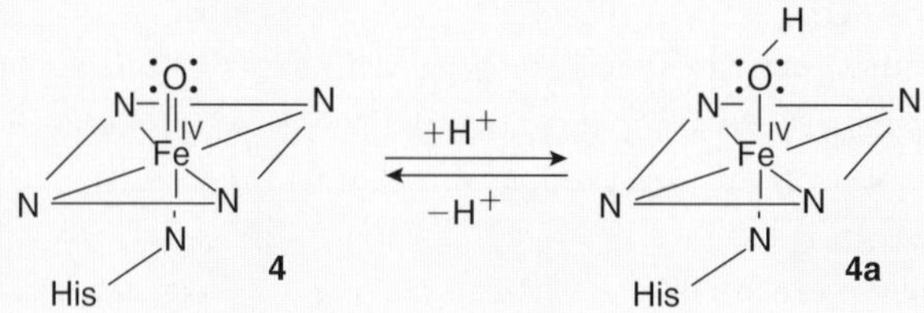

Figure 8. Structures of HRP-II

The formal oxidation state of the iron is IV. The total charge for structure **4** is 0, and for **4a** is +1.

leaving the oxoferryl complex intact. Ortiz de Montellano and co-workers have suggested an alternative, namely that the electron transfer (or hydrogen-atom abstraction) could take place at the edge of the haem rather than at the oxoferryl [Fe(IV)=O] centre [37], and the resulting cation radical of the substrate would disproportionate.

Compound III

HRP-III has a characteristic dark red colour and is equivalent to the oxyferrous complex of HRP, Fe(II)–O_2 or Fe(III)–O_2^- [1]. This complex is also at the same oxidation level and is very similar to oxymyoglobin or oxyhaemoglobin (see Chapter 6 in this volume). The main features in the UV-visible absorption spectrum are the Soret band at 416 nm and two peaks at 546 and 583 nm [1] (Figure 4b). Compound III is not part of the HRP catalytic cycle (and therefore is not shown in the catalytic schemes in this Chapter), but forms upon addition of a large excess of hydrogen peroxide to the native ferric enzyme [1]. It is known as compound III because it is the third complex of HRP to be observed, after compounds I and II, following the addition of hydrogen peroxide. Titration with dimethyl-*p*-phenylenediamine has shown that compound III is three oxidation equivalents above ferric HRP [1], so that it has the formal oxidation state of 6.

Mechanism of formation

HRP-III can be made by three methods. The first involves adding excess hydrogen peroxide to ferric HRP. This reaction proceeds via the generation of compound I and then compound II, which then converts relatively slowly to compound III [1]. Also, the reaction of HRP-II with excess hydrogen peroxide directly yields compound III [38]. A second method involves generation of oxyferrous HRP by addition of molecular oxygen (O_2) to ferrous HRP [1]. Finally, it is reported that compound III can be formed when superoxide anion ($O_2^{\bullet -}$) is added to ferric HRP [1].

Compound 0

The process leading to the formation of compound I in the HRP catalytic cycle is not well understood. Preceding the formation of compound I, HRP interacts with hydrogen peroxide to generate compound 0 (**2**, Figure 1). The many questions that remain concerning the nature of this compound must be addressed to more fully understand the mechanism of compound-I formation.

Baek and Van Wart [39] observed that compound 0 was a transient intermediate *en route* to formation of compound I. At cryogenic temperature, they used rapid-scan spectroscopy to follow the reaction of ferric HRP with hydrogen peroxide or organic hydroperoxides. At ambient temperature, the half-life of the ferric peroxide species (compound 0) has been estimated to range from 0.05 to 0.1 ms, depending on the peroxidase examined. Kinetics and thermodynamics of the formation of compound 0 have also been investigated. Unfortunately, it has not been possible to generate high-enough concentrations of compound 0 for examination with other spectroscopic methods such as Mössbauer spectroscopy, EPR, EXAFS or resonance Raman spectroscopy.

The best evidence for the involvement of compound 0 in formation of compound I comes from kinetic studies. The rate of formation of HRP-I reaches a limiting value at high concentrations of hydrogen peroxide or other alkyl hydroperoxides [39]. This indicates that a precursor to compound I, namely compound 0 (**2**, Figure 1), is formed reversibly *en route* to the generation of compound I (**3**, Figure 1). Similar results were obtained using a water-soluble haem model for peroxidases [40]. In the studies with both HRP and the water-soluble peroxidase model, a transient spectrum was observed that had the surprising features of a "hyperporphyrin" or "split Soret" UV-visible absorption spectrum, with two intense Soret transitions at 400 and 330 nm [40,41]. Such spectra have been observed for both the cytochrome P450 enzymes and CPO [8]. This was the first time such spectra had been observed for a histidine-ligated peroxidase. Thorneley and co-workers carried out rapid-scan stopped-flow studies of compound-I formation in an HRP mutant that had Arg-38 on the distal side replaced by leucine [41]. They observed changes in the Soret absorption bands around 400 nm, but the additional peak at 330 nm was missing.

Harris and Loew [42] have calculated the electronic structure and spectra of resting-state HRP and of two possible forms for compound 0. They concluded that the split Soret UV-visible absorption peaks of compound 0, observed with both wild-type HRP and the water-soluble peroxidase model, represent the ferric anion peroxide complexes [Fe(III)–OOH], whereas the single Soret absorption peak observed with the Arg-38→Leu-mutant HRP was due to the ferric hydrogen peroxide HRP complex, Fe(III)–HOOH [41]. These calculations affirm the conclusions from the above experiments [39–41] that Arg-38 has mechanistic importance in the formation of compound I; these authors suggested that the absence of Arg-38 interferes with proton transfer from hydrogen peroxide to the imidazole of the distal histidine, hindering the O–O heterolytic cleavage and, thus, slowing the rate of compound-I formation [42].

Differences in the catalytic activities of P450 and peroxidases

The reactive-oxygen catalytic species of the cytochrome P450 enzymes and the peroxidases seem to be similar. Both involve oxidation of the ferric resting state to an oxoferryl intermediate and either a porphyrin radical or a protein-based radical. Based on these similarities, the question must be asked, why are the catalytic outcomes of these two systems so very different? Monooxygenases transfer directly the ferryl-bound oxygen to substrate, whereas the peroxidases abstract an electron from their substrates. Unlike the cytochrome P450 enzymes, most peroxidases do not accept electrons from associated electron-transfer proteins and cannot activate dioxygen via the generation of a metal-peroxide species. Cytochrome P450 enzymes can utilize peroxides (peroxygenase rather than oxygenase activity) to support catalysis (no P450 reductase, NADPH or O_2 is required; see Chapter 5 in this volume) while retaining the differences in catalytic mechanism from peroxidases mentioned above.

The reaction of phenylhydrazine with haem proteins such as P450 and myoglobin, in which the haem iron is readily accessible to benzene-ring-bearing substrates or ligands, results in the formation of a phenyl–iron complex [37]. This is not the case for HRP, which undergoes addition of the phenyl group to the porphyrin δ-meso position [37]. Substitution at the haem δ-meso hydrogen with ethyl- or benzene-ring-bearing ligands has been observed to inhibit peroxidase activity in both HRP and CCP [43]. However, replacement of an active-site phenylalanine residue in HRP with smaller amino acids (by site-directed mutagenesis) allows increased *peroxygenase* activity (sulphoxidation of phenyl alkyl thioethers increases by up to 18-fold) and even styrene epoxidation to occur, presumably by altering the protein structure around the haem to increase accessibility to the oxoferryl moiety [44]. Analysis of available peroxidase crystal structures clearly shows the only accessible haem site to be the peripheral δ-meso position [2,5,45]. Thus one difference is in the accessibility of the distal haem site to organic molecules. The peroxidases are hydrophilic and restrict access to the distal side of the haem. This encourages electron-transfer reactions with the porphyrin and interactions with H_2O and hydrogen peroxide. The distal sides of P450 enzymes are more open and hydrophobic, encouraging oxygenations of organic substrates.

The reaction of peracids (e.g. metachloroperbenzoic acid) and alkyl hydroperoxides with P450s and peroxidases gives significantly different results, although compound I forms with both. Whereas the haem in cytochrome P450 is totally destroyed in seconds, the haem in HRP is only occasionally modified [32], and the haem in CPO is not damaged to any significant extent [26]. The results of such studies support the notion that a major reason the peroxygenase activity of typical peroxidases is suppressed is because of the steric and polar constraints of the protein, which limits access of organic substrates to the ferryl oxygen [37]. Perhaps this is the principle reason why cytochrome P450 enzymes and peroxidases behave so differently.

Perspectives

This Chapter has focused on the different peroxidase states, including their structures, how they can be formed, and some of the reactions they can undergo. Using various spectroscopic techniques and site-directed mutagenesis, many discoveries have been made to elucidate the sequence of reactions and the reactive intermediates involved.

In the case of HRP and other histidine-ligated peroxidases, the final piece of the puzzle seems to be the crystal structure. The crystal structures of the numerous enzymes and some of their derivatives have confirmed many of the suggestions put forth for the structures of these peroxidases, while refuting others. These crystal structures also clarified their catalytic nature by identifying the individual amino acid residues involved in the reaction sequence. Such knowledge is invaluable for protein engineering that aims to construct various types of enzyme to service many specific tasks.

On the other hand, in the case of CPO, more work is still needed. CPO is unique among peroxidases. Whereas other peroxidases have a histidine proximal haem-iron ligand, CPO employs cysteineate. Such proximal ligation has earned CPO special recognition because it is similar to cytochrome P450. This resemblance implies similar chemical properties and similar reactive intermediates, and has been the basis for CPO being used as a reasonable protein model for the study of P450. Determination of the properties of the high-valency intermediates in the CPO reaction cycle has proven to be more elusive than with other peroxidases such as HRP. Stopped-flow spectroscopic techniques (including rapid-scan methods) will be required to obtain the appropriate kinetics data for delineating intermediates in the reactions. Rapid freeze–quench methodologies can then be used to capture the elusive CPO-intermediate species for further characterization by magnetic-resonance techniques. This will enable determination of the redox and co-ordination states of the intermediates.

The potential for using haem-containing peroxidases and other haem-containing enzymes in bioremediation processes in the environment is enormous. For that specific reason, better understanding of the structure–function relationships of peroxidase enzymes is necessary for constructing specifically engineered peroxidases to accomplish many of these difficult and otherwise unfeasible tasks.

Summary

- *Peroxidases are enzymes that utilize hydrogen peroxide to oxidize substrates.*
- *A histidine residue on the proximal side of the haem iron ligates most peroxidases.*
- *The various oxidation states and ligand complexes have been spectroscopically characterized.*
- *HRP-I is two oxidation states above ferric HRP. It contains an oxoferryl (=oxyferryl) iron with a π-radical cation that resides on the haem. HRP-II is one oxidation state above ferric HRP and contains an oxoferryl iron. HRP-III is equivalent to the oxyferrous state.*

- *Only compounds I and II are part of the peroxidase reaction cycle.*
- *CCP-ES contains an oxoferryl iron but the radical cation resides on the Trp-191 residue and not on the haem.*
- *CPO is the only known peroxidase that is ligated by a cysteine residue rather than a histidine residue, on the proximal side of the haem iron. CPO is a more versatile enzyme, catalysing numerous types of reaction: peroxidase, catalase and halogenation reactions.*
- *The various CPO species are less stable than other peroxidase species and more elusive, thus needing further characterization.*
- *The roles of the amino acid residues on the proximal and distal sides of the haem need more investigation to further decipher their specific roles.*
- *Haem proteins, especially peroxidases, are structure–function-specific.*

The National Science Foundation and the National Institutes of Health have supported research in the Dawson laboratory on peroxidase enzymes. We would like to thank Dr. Alycen Pond and Dr. Masanori Sono and Dr. David Ballou for helpful discussions.

References

1. Dunford, H.B. & Stillman, J.S. (1976) On the function and mechanism of action of peroxidases. *Coord. Chem. Rev.* **19**, 187–251
2. Gajhede, M., Schuller, D.J., Henriksen, A., Smith, A.T. & Poulos, T.L. (1997) Crystal structure of horseradish peroxidase C at 2.15 Å resolution. *Nat. Struct. Biol.* **4**, 1032–1038
3. English, A.M. & Tsaprailis, G. (1995) Catalytic structure-function relationships. Heme peroxidases. *Adv. Inorg. Chem.* **43**, 79–125
4. Sundaramoorthy, M., Terner, J. & Poulos, T.L. (1995) The crystal structure of chloroperoxidase: A heme peroxidase-cytochrome P450 function hybrid. *Structure* **3**, 1367–1377
5. Henriksen, A., Welinder, K. & Gajhede, M. (1998) Structure of barley grain peroxidase refined at 1.9 Å resolution. *J. Biol. Chem.* **273**, 2241–2248
6. Fülöp, V., Ridout, C.J., Greenwood, C. & Hajdu, J. (1995) Crystal structure of the di-heme cytochrome *c* peroxidase from *Pseudomonas aeruginosa*. *Structure* 3, 1225–1233
7. Marnett, L.J. & Kennedy, T.A. (1995) Comparison of the peroxidase activity of hemeproteins and cytochrome P450. In *Cytochrome P450: Structure, Mechanism, and Biochemistry*, 2nd edn. (Ortiz de Montellano, P.R., ed.), pp. 49–80, Plenum, New York
8. Dawson, J.H. & Sono, M. (1987) Cytochrome P-450 and chloroperoxidase: thiolate-ligand heme enzymes. Spectroscopic determination of their active site structures and mechanistic implications of thiolate ligation. *Chem. Rev.* **87**, 1255–1276
9. Kaneko, Y., Tamura, M. & Yamazaki, I. (1980) Formation of porphyrin π-cation radical in zinc-substituted horseradish peroxidase. *Biochemistry* **19**, 5795–5799
10. Penner-Hahn, J., McMurry, T.J., Renner, M., Latos-Grazynsky, L., Eble, K.S., Davis, I.M., Balch, A.L., Groves, J.T., Dawson, J.H. & Hodgson, K.O. (1983) X-ray absorption spectroscopic studies of high valent iron porphyrins. Horseradish peroxidase compounds I and II and synthetic models. *J. Biol. Chem.* **258**, 12761–12764
11. Schulz, C.E., Devaney, P.W., Winkler, H., Debrunner, P.G., Doan, N., Chang, R., Rutter, R. & Hager, L.P. (1979) Horseradish peroxidase compound I: evidence for spin coupling between the heme iron and a "free" radical. *FEBS Lett.* **103**, 102–105

12. Penner-Hahn, J., Eble, K.S., McMurry, T.J., Renner, M., Balch, A.L., Groves, J.T., Dawson, J.H. & Hodgson, K.O. (1986) Structural characterization of horseradish peroxidase using EXAFS spectroscopy. Evidence for Fe=O ligation in compounds I and II. *J. Am. Chem. Soc.* **108**, 7819–7825
13. Roberts, J.E., Hoffman, B.M., Rutter, R. & Hager, L.P. (1981) Electron-nuclear double resonance of horseradish peroxidase compound I. Detection of the porphyrin π-cation radical. *J. Biol. Chem.* **256**, 2118–2121
14. Roberts, J.E., Hoffman, B.M., Rutter, R. & Hager, L.P. (1981) ^{17}O ENDOR of horseradish peroxidase compound I. *J. Am. Chem. Soc.* **103**, 7654–7656
15. Palaniappa, V. & Terner, J. (1989) Resonance Raman spectroscopy of horseradish peroxidase derivatives and intermediates with excitation in the near ultraviolet. *J. Biol. Chem.* **264**, 16046–16053
16. La Mar, G., de Ropp, J.S., Smith, K.C. & Langry, K.C. (1981) Proton nuclear magnetic investigation of the electronic structure of compound I of horseradish peroxidase. *J. Biol. Chem.* **256**, 237–243
17. Hanson, L.K., Chang, C.K., Davies, M.S. & Fajer, J. (1981) Electron pathways in catalase and peroxidase enzymic catalysis. Metal and macrocycle oxidations of iron porphyrins and chlorins. *J. Am. Chem. Soc.* **103**, 663–670
18. Schonbaum, G.R. & Lo, S. (1972) Interaction of peroxidases with aromatic peracids and alkyl peroxides. *J. Biol. Chem.* **247**, 3353–3360
19. Edwards, S.L., Nguyen, H.X., Hamlin, R.C. & Kraut, J. (1987) Crystal structure of cytochrome *c* peroxidase compound I. *Biochemistry* **26**, 1503–1511
20. Pond, A.E., Bruce, G.S., English, A.M., Sono, M. & Dawson, J.H. (1998) Spectroscopic study of the compound ES and the oxoferryl compound II states of cytochrome *c* peroxidase: comparison with compound II of horseradish peroxidase. *Inorg. Chim. Acta* **275–276**, 250–255
21. Chance, M., Powers, L., Poulos, T.L. & Chance, B. (1986) Cytochrome *c* peroxidase compound ES is identical with horseradish peroxidase compound I in iron-ligand distances. *Biochemistry* **25**, 1266–1270
22. Hashimoto, S., Teraoka, J., Inubush, T., Yonetani, T. & Kitagawa, T. (1986) Resonance Raman study on cytochrome *c* peroxidase and its intermediates. *J. Biol. Chem.* **261**, 11110–11118
23. Yonetani, T., Schleyer, H. & Ehrenberg, A. (1966) Studies on cytochrome *c* peroxidase. *J. Biol. Chem.* **241**, 3240–3243
24. Sivararja, M., Goodin, D.B., Smith, M. & Hoffman, B.M. (1989) Identification by ENDOR of Trp^{191} as the free-radical site in cytochrome *c* peroxidase. *Science* **245**, 738–740
25. Huyett, J.E., Doan, P.E., Gurbiel, R., Houseman, A.L.P., Sivaraja, M., Goodin, D.B. & Hoffman, B.M. (1995) Compound ES of cytochrome *c* peroxidase contains a Trp π-cation radical: characterization by CW and pulsed Q-band ENDOR spectroscopy. *J. Am. Chem. Soc.* **117**, 9033–9041
26. Isaac, I.S. (1998) Mechanistic and structural insights from the reactions of peroxidases and cytochrome P450cam with metachloroperbenzoic acid. Ph.D. Thesis, University of South Carolina
27. Choudhury, K., Sundaramoorthy, M., Hickman, A., Yonetani, T., Woehl, E., Dunn, M.F. & Poulos, T.L. (1994) Role of the proximal ligand in peroxidase catalysis. *J. Biol. Chem.* **269**, 20239–20249
28. Urano, Y., Higuchi, T., Hirobe, M. & Nagano, T. (1997) Pronounced axial thiolate ligand effect on the reactivity of high-valent oxo-iron porphyrin intermediate. *J. Am. Chem. Soc.* **119**, 12008–12009
29. Champion, P.M. (1989) Elementary electronic excitations and the mechanism of cytochrome P450. *J. Am. Chem. Soc.* **111**, 3433–3434
30. Loew, G.H., Collins, J., Luke, B., Waleh, A. & Pudzianowski, A. (1986) Theoretical studies of cytochrome P-450. Characterization of stable transient active states, reaction mechanisms and substrate-enzyme interactions. *Enzyme* **36**, 54–78
31. Chamulitrat, W., Takahashi, N. & Mason, R.P. (1989) Peroxyl, alkoxyl, and carbon-centered radical formation from organic hydroperoxides by chloroperoxidase. *J. Biol Chem.* **264**, 7889–7899
32. Rodriguez-Lopez, J.N., Herñandez-Ruiz, J., Carciu-Canovas, F., Thorneley, R.N.F., Acosta, M. & Arnao, M. (1997) The inactivation and catalytic pathways of horseradish peroxidase with *m*-chloroperoxybenzoic acid. *J. Biol. Chem.* **272**, 5469–5476

33. Hayashi, Y. & Yamazaki, I. (1979) The oxidation-reduction potentials of compound I/compound II and compound II/ ferric couples of horseradish peroxidase A_2 and C. *J. Biol. Chem.* **254**, 9101–9106
34. Chance, B., Powers, L., Ching, Y., Poulos, T., Schonbaum, G.R., Yamazaki, I. & Paul, K.G. (1984) X-ray absorption studies of intermediates in peroxidase activity. *Arch. Biochem. Biophys.* **235**, 596–611
35. Sitter, A.J., Reczek, C.M. & Terner, J. (1985) Heme-linked ionization of horseradish peroxidase compound II monitored by the resonance Raman Fe(IV)=O stretching vibration. *J. Biol. Chem.* **260**, 7515–7522
36. Yamada, H. & Yamazaki, I. (1974) Proton balance in conversions between five oxidation-reduction states of horseradish peroxidase. *Arch. Biochem. Biophys.* **165**, 728–738
37. Ator, M., David, S.K. & Ortiz de Montellano, P.R. (1989) Stabilized isoporphyrin intermediates in the inactivation of horseradish peroxidase by alkylhydrazines. *J. Biol. Chem.* **264**, 9250–9257
38. Nakajima, R. & Yamazaki, I. (1987) The mechanism of oxyperoxidase formation from ferryl peroxidase and hydrogen peroxide. *J. Biol. Chem.* **262**, 2576–2581
39. Baek, H.K. & Van Wart, H.E. (1992) Elementary steps in the reaction of horseradish peroxidase with several peroxides: kinetics and thermodynamics of formation of compound 0 and compound I. *J. Am. Chem. Soc.* **114**, 718–725
40. Wang, J.S., Baek, H.K. & Van Wart, H.E. (1991) High-valent intermediates in the reaction of N-alpha-acetyl microperoxidase-8 with hydrogen peroxide: models for compounds 0, I and II of horseradish peroxidase. *Biochem. Biophys. Res. Commun.* **179**, 1320–1324
41. Rodriguez-Lopez, J.N., Smith, A.T. & Thorneley, R.N.F. (1996) Role of arginine 38 in horseradish peroxidase. *J. Biol. Chem.* **271**, 4023–4030
42. Harris, D.L. & Loew, G.H. (1996) Identification of putative peroxide intermediates of peroxidases by electronic structure and spectra calculations. *J. Am. Chem. Soc.* **118**, 10588–10594
43. Song, W.-C., Funk, C.D. & Brash, A.R. (1993) Molecular cloning of an allene oxide synthase: a cytochrome P450 specialized for the metabolism of fatty acid hydroperoxides. *Proc. Natl. Acad. Sci. U.S.A.* **90**, 8519–8523
44. DePillis, G.D., Sishta, B.P., Mauk, A.G. & Ortiz de Montellano, P.R. (1991) Small substrates and cytochrome *c* are oxidized at different sites of cytochrome *c* peroxidase. *J. Biol. Chem.* **266**, 19334–19341
45. Ozaki, S.I. & Ortiz de Montellano, P.R. (1994) Molecular engineering of horseradish peroxidase. Highly enantioselective sulphoxidation of aryl alkyl sulphides by Phe-41→Leu mutant. *J. Am. Chem. Soc.* **116**, 4487–4488

5

Nature's universal oxygenases: the cytochromes P450

Stephen G. Sligar

Beckman Institute, University of Illinois, 405 N. Mathews Ave., Urbana, IL 61801, U.S.A.

Introduction

Faced with a plethora of recalcitrant hydrocarbon compounds, Nature needed a means for catalytically functionalizing these structures as initial steps in catabolic pathways leading to smaller or more polar metabolites. A separate challenge for evolutionary development was to devise a means of derivatizing a single precursor molecule (e.g. cholesterol) into several unique regio- and stereochemically defined structures that can be used as signals for specifically regulating a variety of cellular metabolic and differentiation processes. Such problems face all forms of life, from humans and other mammals to the great diversity of plants, animals, insects and microbes. One solution to this problem that is found in Nature is the P450 mono-oxygenases, one of the most varied and largest families of metalloenzymes [1].

In biochemistry we often name proteins that have functions that are as important to life as cytochrome P450 simply with the letter 'P' followed by a number. There are many 'p's' or 'P's' in biochemistry, sometimes referring to the size of a protein (often a band characterized by electrophoretic mobility, such as the cancer-linked oncogene p21, for 21 kDa) or to a pigment with the number following the 'P' referring to the wavelength of maximal optical absorption. Examples include the P700 and P680 systems of photosynthesis and the P450 mono-oxygenases. In this latter case, 450 refers to the wavelength of maximal absorbance observed when a sample of ground rat liver is reduced by dithionite and reacted with carbon monoxide. This pioneering discovery by Garfinkel [2] and Klingenberg [3] in the late 1950s was the result of their investigation into how the liver metabolized extra-corporeally administered drugs. Previously, carbon monoxide had been used to probe for haem proteins that could form stable carbonyl adducts in the ferrous iron state. It is likely that these investigators were looking for a pigment that, when reduced and treated with carbon monoxide, absorbed light at 420 nm, like the oxygen-carrying proteins haemoglobin and myoglobin, or possibly light that is slightly longer in wavelength, in the neighbourhood of 430 nm. Thus an absorbance at 450 nm was novel and unexpected.

We now know, in exquisite detail, why the absorbance of the carbon monoxide complex of this reduced haem protein occurs at 450 nm. In this case, the Soret band (the most intensely absorbing band of haems) is split from a roughly 420-nm centroid to bands at ≈370 and ≈450 nm. This was most clearly shown by Hanson et al. in the 1970s using single-crystal optical spectroscopy and extended Hückle calculations [4]. Mason et al. [5] first proposed that this unique 450-nm spectral signature was caused by ligation of a protein cysteine thiolate to the Fe^{2+} of the haem. This hypothesis was later confirmed by Champion et al. [6], who carried out elegant Raman-spectroscopic experiments using isotopically substituted samples, and by Poulos et al. [7], who used X-ray crystallography. Comparative spectroscopic studies with chemically defined model systems by Dawson and others also supported this conclusion [8].

Historically, there were questions about the function of this unique haem protein. Pioneering experiments by Cooper et al. [9] positively identified this liver enzyme as being the major player in xenobiotic transformations by using the action spectrum for photolytically reversing the inhibition of drug metabolism caused by carbon monoxide. In these studies, it was found that the carbon monoxide inhibition of drug metabolism was reversed most effectively by irradiating with light corresponding to the peaks of the P450 absorbance spectrum. Heroic efforts by Lu and Coon led to a procedure for purifying this membrane protein to homogeneity and in high yield, enabling characterization by the precise methods of physical biochemistry [10]. Parallel and independent studies in the late 1960s by those in the Gunsalus laboratory, who were studying terpene metabolism in *Pseudomonas* bacteria, made it possible to purify easily a soluble stable P450 cytochrome for mechanistic investigations [11].

Classification of P450s

The cytochromes P450 enjoy significant representation across a wide diversity of life-forms. Scientists tend to classify things, so one can ask what ways are there to group the cytochrome P450 enzymes to show several fundamental linkages of evolutionary function. Figure 1 illustrates one representation of the 'tree of life' that attempts to represent the present biological diversity through

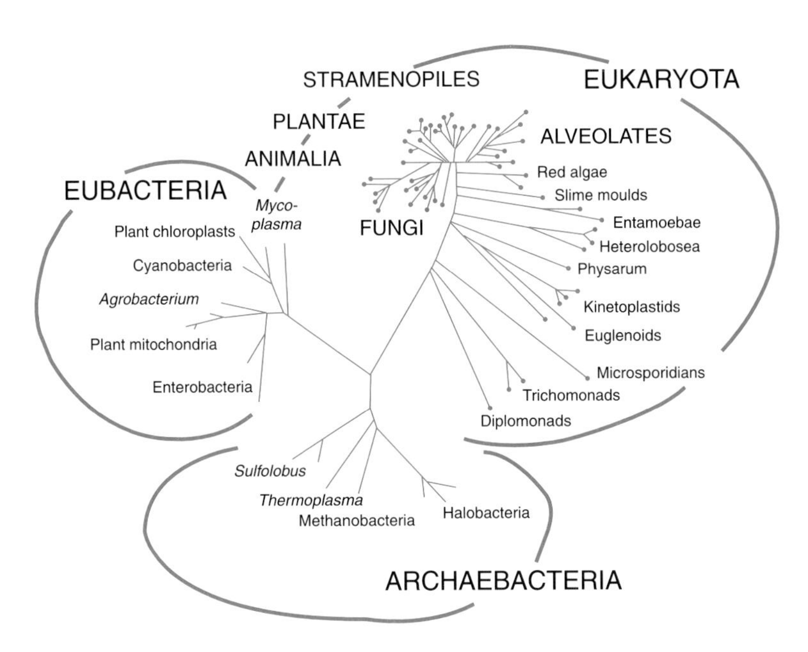

Figure 1. Evolutionary 'tree of life' based on 16 S ribosomal RNA sequences
Adapted from [12] with permission. ©1997 Cell Press.

genetic evolution from primordial ancestors [12]. This particular tree is borrowed from the fundamental genomic work of Woese and co-workers at the University of Illinois, who discovered the Archaea by studying the evolution of 16 S ribosomal RNA [13]. It is far beyond the scope of this review to discuss the field of genomics, except to note that the cytochrome P450 systems offer an ideal group of enzymes for bioinformatic investigation. There are currently more than 700 P450 gene sequences known, 500+ of which have been carefully aligned and documented by D. Nelson on his very useful and informative website for the Committee for Standardized Cytochrome P450 Nomenclature (http://drnelson.utmem.edu/nelsonhomepage.html). Sequence comparisons divide the P450 sequences into clans of families and variants. The linkage of diverse sequences that share common chemical function, but have weak primary homology, is an important and hotly pursued topic in the field of genomics. Proteomics, the coupling of function across evolution, is clearly a very exciting area of P450 investigation. It will be interesting to see if a 'tree of life' constructed from P450 sequences will be similar to that shown in Figure 1, which is based on 16 S RNA sequences.

The cytochromes P450 can be grouped according to their metabolic functions. Many P450s are involved in the reactions of catabolism. In humans, the introduction of compounds that originate from outside the body — so-called xenobiotics — triggers the induction of P450 enzymes in the lung, liver and epithelial tissues; these P450s break down (or make more hydrophilic and water soluble) xenobiotics (such as drugs) to structures more appropriate for their excretion. The wide variety of pharmaceutical agents currently being utilized by humans has made the study of drug metabolism important. Because the P450 family of enzymes participates in the metabolism of a large fraction of all drugs used in medicine, P450 enzymes are often the major limitation of their usefulness. This property is, in fact, a major problem for the development of drugs that will remain in humans long enough to be effective. Because the mean plasma lifetime of nearly all therapeutic agents is related to the activity of the cytochromes P450 in liver, many investigations into this ubiquitous metalloprotein involve the study of hepatic microsomal fractions. The liver endoplasmic reticulum is the site of several dozen P450 isozymes; some are constitutively expressed, others are induced by the action of specific compounds, but each exhibits a relatively broad range of substrate specificity. One isozyme may preferentially hydroxylate aromatic ring structures, another may prefer to oxidatively demethylate amines, whereas yet another may be responsible for metabolizing alcohols, and so on. These catabolic P450s provide important defence mechanisms for humans; thus xenobiotics are degraded, and hydrophobic compounds are solubilized so that they do not sequester in fat tissues where they would be recalcitrant to elimination. However, the distribution of P450s in individuals can vary considerably. Thus the usefulness of particular drugs or combinations of drugs may also vary from person to person. This constitutes a major problem in the treatment of illness.

A different group of the cytochromes P450 is involved in equally important anabolic metabolic transformations, such as several of the steps of hormone biosynthesis. In humans, all steroid hormones are made from cholesterol through metabolic transformations occurring in the adrenal glands, ovaries and testes. The first major step involves three separate oxidation reactions that cleave the isocaproic aldehyde from the apex of the D-ring to form pregnenolone. Separate P450 enzymes then generate the various classes of androgens and oestrogens by highly regio-selective and stereospecific hydroxylations. In contrast to the catabolic role of hepatic P450 systems, which require broad substrate tolerance, each of the P450s involved in hormone biosynthesis is finely tuned to act on only a single substrate to make a particular product. The body would not want to confuse androgens and oestrogens! Similar, highly controlled, P450-catalysed metabolic transformations are involved in the synthesis of other very active biological signalling agents, such as the prostaglandins, which are derived from unsaturated C_{20} fatty acids. Like the drug-metabolizing enzymes, the P450s participating in hormone biosynthesis are membrane-associated, often in the mitochondrial fraction of the cell. Due to the exquisite sensitivity of organisms to small quantities of hormones, this class of P450s is usually very highly regulated at the transcription level.

As one looks across the 'tree of life' in Figure 1, a similar division of P450 metabolic functions seems to apply. P450s are found in all three major branches of the evolutionary tree. Insects, for example, utilize P450 in a catabolic role to degrade insecticides in the fat body, rendering protection from chemical attack. At the same time, they utilize P450s in an anabolic reaction to control

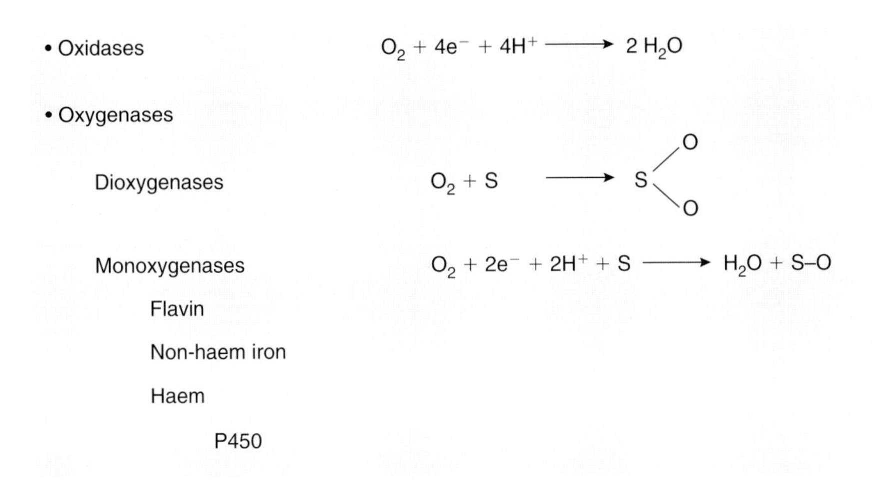

Figure 2. The enzyme systems of oxygen metabolism

hormone synthesis, such as the 20-α hydroxylation of ecdysone to form ecdasterone, which initiates moulting in the fifth instar of development. Plants have analogous transformations, utilizing P450s to degrade herbicides for plant protection as well as to synthesize a variety of complicated plant hormones via highly specific oxygenase chemistry.

The enzymes of oxygen metabolism are described in Figure 2. The oxidases utilize atmospheric dioxygen as a terminal oxidant, reductively cleaving the oxygen–oxygen bond with four electrons and four protons to yield two water molecules as part of the energy-generating pathways of aerobic organisms. The cytochromes P450 are members of the oxygenase class of enzymes. These enzymes also act on atmospheric dioxygen, but involve a co-oxidation of a hydrocarbon substrate. The dioxygenases insert both atoms from atmospheric O_2 into the carbon chain of a substrate, whereas the mono-oxygenases incorporate a single oxygen atom, reducing the other atom to a molecule of water with two electrons and two protons (see Chapters 3 and 11 in this volume). Mono-oxygenases can employ any of a variety of prosthetic groups, including flavins, single or multiply co-ordinated transition metals, or haems. The P450s belong to this latter group of metalloenzymes. In all cases, the two reducing equivalents required for the P450 stoichiometry are provided by a reduced pyridine nucleotide, either NADH or NADPH.

Cytochromes P450 can also be classified by the type of redox-transfer chain that is used to bring the needed electrons into the haem active centre; see Figure 3. Class-I P450s utilize a flavoprotein dehydrogenase that contains an FAD prosthetic group. A hydride is transferred from NAD(P)H to the FAD, and then the two-electron equivalents are transferred sequentially to low-molecular-mass iron–sulphur redoxins. This transforms the two-electron processes normally involved in metabolic transformations to one-electron transfer chains. The redoxins contain two antiferromagnetically coupled iron atoms

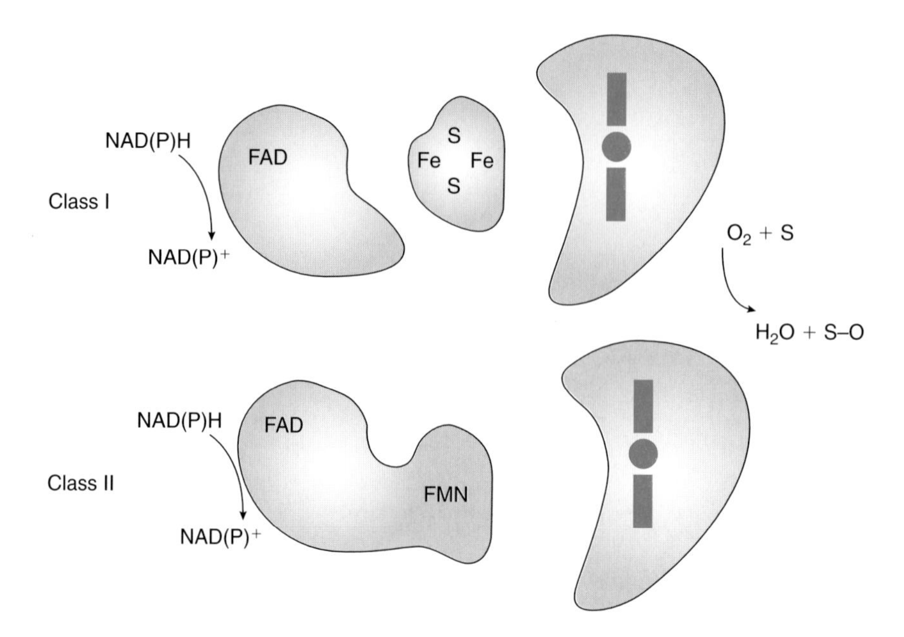

Figure 3. Classification of the cytochromes P450 by the nature of their redox-transfer systems

bridged by inorganic sulphides (see Chapters 3 and 7 in this volume). These one-electron redox carriers donate electrons into the P450 haem centre in two separate steps of the oxygenation reaction (see below). The class-II P450s utilize a single more-complicated flavoprotein reductase that carries out both of the above functions. It similarly uses an FAD centre to accept a hydride from NADPH, but substitutes an FMN prosthetic group for the iron–sulphur protein. This FAD/FMN diflavin reductase makes use of all three redox states of the isoalloxazine ring system of the flavins; reduced, semi-quinone and oxidized states, in serving as the hydride-to-one-electron transformer that feeds into the P450 haem centre.

The P450 haem proteins are of molecular mass ≈50000 Da and house the catalytic site that binds substrate and atmospheric dioxygen. Electron input from the associated redox-transfer chain initiates the reductive scission of the dioxygen bond of O_2, the production of water and the oxygenation of the substrate. P450s can carry out a wide variety of chemical transformations, as indicated in Figure 4. Perhaps most interesting from the chemical point of view is the hydroxylation of unactivated hydrocarbons to form products such as primary alcohols. This is the only mammalian enzyme system with the oxidative power to effectively functionalize primary carbons using molecular oxygen.

Mechanism

Due to the enormous diversity of the P450s and their important roles in a wide variety of metabolic transformations across all life-forms, their chemical

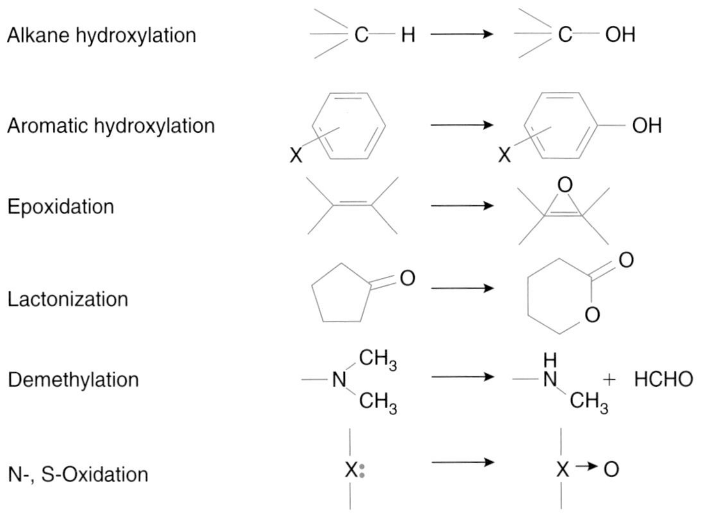

Alkane hydroxylation

Aromatic hydroxylation

Epoxidation

Lactonization

Demethylation

N-, S-Oxidation

Figure 4. Chemical transformations, oxygen-insertion reactions, carried out by the cytochromes P450

mechanisms have received considerable attention from pharmacologists, toxicologists, chemists, physicists and molecular biologists. Figure 5 presents a current view of the chemical reaction cycle of the cytochromes P450, and illustrates schematically some of the interesting molecular questions that are associated with each elementary step of the cycle [14].

Beginning with the ferric haem resting state **1** (see Figure 5), which usually contains a haem-bound water as sixth axial ligand, the P450 binds the substrate. This binding reaction utilizes the features of the protein active site, which in some forms of P450 can convey a high degree of substrate selectivity needed for function *in vivo*. In some cases, a high degree of substrate–pocket complementarity displaces the bound water, causing the *d*-electron configuration of the haem iron to change from a low-spin to a high-spin ferric state (species **2**) with a concomitant increase in the redox potential of the haem. In the class-I enzymes, the intermediate iron–sulphur proteins that provide the reducing equivalents required for dioxygen cleavage have a redox potential near −230 mV. The binding of a cognate substrate brings the haem redox potential to a value higher than that of the iron–sulphur protein, so that this binding process serves as a form of molecular switch to permit electron transfer into the haem centre. The first redox equivalent transferred is responsible for a ferric–ferrous change in the haem (species **3**). Many exciting questions at the centre of biophysics and bioinorganic chemistry deal with the precise mechanisms of inter- and intra-protein electron transfer, such as are exemplified in the P450 systems.

Continuing along the reaction cycle, the ferrous haem binds dioxygen to form **4** in a reaction analogous to that of the well-studied oxygen storage and

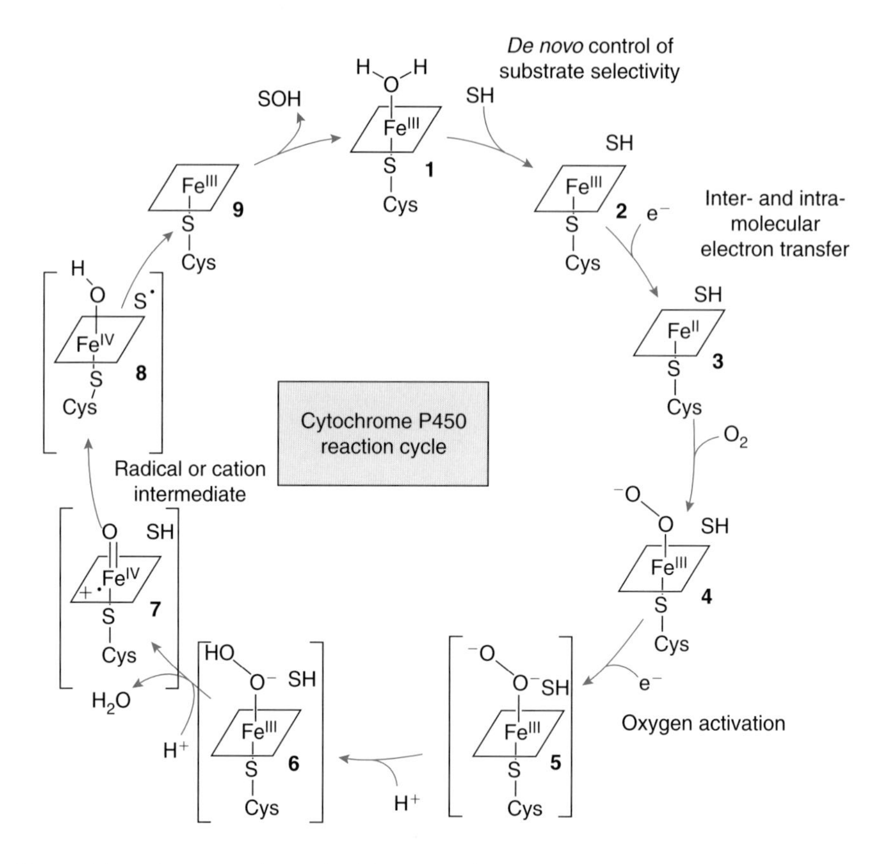

Figure 5. Unsolved questions of P450 mechanism

transport proteins, haemoglobin and myoglobin (see Chapter 6 in this volume). This reaction of the P450s has not been explored extensively. Given the newly realized roles for diatomic gaseous ligands such as nitric oxide and carbon monoxide in vasoconstriction and signalling, the mechanisms for the control of rate and selectivity with O_2 binding to the P450 haem pocket need further attention.

Oxygenated haemoproteins have varying degrees of stability against the release of superoxide that occurs with concomitant formation of the ferric resting state of the haem, a process known as autoxidation. The globins, which function to reversibly bind and transport oxygen, are perhaps the most stable, with autoxidation measured in hours or days. The P450s autoxidize much faster, on the order of minutes or seconds. This process has been well studied using cryo-enzymology for the bacterial $P450_{cam}$ from *Pseudomonas*, whose oxygenated haem complex autoxidizes at a rate of $4 \times 10^{-3} s^{-1}$ [15]. What makes P450 unique from the globins, however, is the ability to efficiently take up a second reducing equivalent from the associated electron-transport chain, and thereby reductively cleave the oxygen–oxygen bond. This produces a single water molecule and a high-valency metal–oxo complex that is competent in substrate oxygenation.

The intermediates shown bracketed in Figure 5 (**5**–**8**) are hypothetical, and their support comes primarily from stoichiometric factors and from comparisons with other metalloporphyrin enzymes and model systems (see Chapter 4 in this volume). For the sake of discussion, we show the first bracketed species **5** as a two-electron-reduced, dioxygen-bound, haem adduct; rupture of the oxygen–oxygen bond, accompanied by controlled proton delivery (perhaps giving species **6**, which releases water), leaves a single oxygen atom bound to the transition metal **7**. This state is formally two oxidation equivalents above the ferric haem resting state and is analogous to the compound-I intermediate of peroxidases as discussed by Isaac and Dawson in Chapter 4 in this volume. Such a species is predicted to be strongly oxidizing. In contrast to the peroxidases, however, this intermediate has never been directly observed in an oxygen-driven P450 reaction. In the late 1970s, Groves postulated that such an 'oxene' intermediate could oxygenate a substrate through two-steps; hydrogen abstraction followed by very rapid radical recombination of the resulting iron-bound hydroxyl radical, in a process they termed 'oxygen rebound' [16]. The strong evidence that is compatible with this radical-based mechanism is the widely observed racemization of the stereochemistry at the carbon that is oxidized. More recently, there have been other suggestions for chemical processes that utilize higher-valency metal–oxo complexes in carbon-chain functionalization, including the 'agostic' hypothesis of Collman et al. [17], and a reactive iron-bridged carbon–oxygen species postulated by Yoshizawa [18].

Interestingly, pioneering work from the Coon and Vaz laboratories [19] has shown that other intermediates in the reaction cycle of cytochrome P450 can also effectively carry out some of the diverse transformations shown in Figure 4. Examples include the distal nucleophilic attack by the iron–peroxo species **5** (a precursor to the iron–oxo compound) as the key step in reactions of aldehyde species in the aromatase reactions of steroid biosynthesis, or the side-chain cleavage of arginine in the analogous nitric oxide synthase. Newcomb et al. [20] suggested that a protonated peroxide complex **6** (the hydroperoxide shown in Figure 5) could act as an electrophile in some epoxidation reactions. Clearly, it is the enormous versatility of this enzyme in terms of both substrate specificity (or tolerance) and chemical catalysis that has made it such a central player in the wide variety of life-forms indicated in Figure 1.

An in-depth discussion of all of the interesting unsolved questions of P450 catalysis indicated in Figure 5 is clearly beyond the scope of this Chapter. However, numerous recent reviews have appeared, and a major monograph edited by Ortiz de Montellano perhaps is readily accessible [21]. In addition, another useful website, the Directory of P450-Containing Systems, for the latest information about P450-related systems, is available at http://www.icgeb.trieste.it/p450/. One of the more puzzling questions involves the role of the unique cysteine axial ligand in providing special chemical reactivity to the metal centre in one or more of the fundamental stages of the P450 reaction cycle. Roles that have been suggested include: the thiolate axial ligand is responsible for the low redox potential of the prosthetic group; its basic character is critical for the oxygen–oxygen bond scission event; and it fine-tunes and permits radical reactivity for the putative higher-valency metal–oxo complex. It is probably fair to say at present that there is no unambiguous documentation of why Nature utilizes this relatively rare amino acid to co-ordinate the metal centre of the cytochromes P450. Perhaps it was simply to give the early investigators the spectroscopic handle to find this important enzyme in crude organ homogenates!

One of the most intense focuses of current chemical research in the P450 arena is in elucidating the detailed mechanism for transformation of the ferrous dioxygen-bound intermediate **4** into the iron–oxo haem **7** with concomitant release of water. This oxygen–oxygen bond cleavage step is central to the unique chemical events involved in oxygenation of unactivated carbon chains by the P450s. The key chemical steps occur after the transfer of the second electron from the redox partner to the haem (**4**→**5**) and the associated proton movement (**5**→**6**). There has been an intense interest in the physics and chemistry communities for several years about the nature of inter-protein and intra-protein electron transfer. Recently, two schools of thought have emerged. One, championed by Gray and Winkler [22], suggests that redox movement inside proteins occurs by a super-exchange mechanism, and that Nature has 'wired' efficient electron-transfer pathways in proteins by taking advantage of the electronic structure of covalent bonds, hydrogen bonds, and the efficiencies of through-space quantum-mechanics tunnelling. An alternate view, advocated by Moser and Dutton [23], also involves a super-exchange mechanism for electron transfer, but notes that there is a general lack of specificity in the control of redox movement beyond a simple dependence on the distance between donor and acceptor sites. A more detailed discussion may be found in Chapter 7 of this volume. In looking at the redox-linked dioxygen activation step shown in Figure 5, one must remember that Nature repeatedly uses one of two fundamental chemical processes; acid/base catalysis or nucleophilic catalysis. It is far easier to control acid/base reactivity at an enzyme active site than a detailed pathway of electron transfer. For this reason my laboratory has been exploring the details of dioxygen reactivity with protein-bound haems by looking at key steps of proton delivery that might be linked to oxygen–oxygen bond scission.

A prevalent method for unravelling the mysteries of enzyme catalysis since the 1980s has been the use of site-directed mutagenesis to alter specific side chains that are thought to be chemically involved in an active site. The pioneering work of Poulos et al. led to the X-ray structure of a $P450_{cam}$ from *Pseudomonas* in 1985 (described in [7]). Many expected the active site to comprise the type of acid/base functionalities that were observed previously in the structures of the peroxidases. There, as discussed in Chapter 4 of this volume, arginine and histidine residues are precisely positioned to remove a proton from incoming hydrogen peroxide and to aid in the heterolytic scission of the O–O bond. The P450s, however, operate very differently from the peroxidases, in that the peroxide oxidation state is generated from the addition of electrons to an unprotonated, but co-ordinated, dioxygen, rather than from the addition of H_2O_2 to the ferric haem. Thus the pK values of the corresponding transformations in the two types of protein are quite different. Through site-directed mutagenesis of acidic and alcoholic amino acids, which are highly conserved in the active site of P450$_{cam}$ and most known P450 sequences, my laboratory and that of Ishimura in Japan proposed the mechanism of proton delivery and subsequent dioxygen-bond cleavage that is illustrated in Figure 6 [14]. Thr-252 (in the P450$_{cam}$ nomenclature) plays a key role in stabilizing the end-on co-ordination of bound dioxygen through a bridging water molecule. In P450$_{cam}$, which is optimized for O–O bond scission rather than for using the distal oxygen atom of the bound O_2 as a nucleophile, this mode of co-ordination that is stabilized by Thr-252 also helps to avoid the unwanted release of hydrogen peroxide following input of the second electron. The key process of O–O bond scission is envisioned to occur by selective proton delivery through bridging waters. The participation of an aspartic acid side chain (Asp-251) serves as a 'switch' to shuttle protons from a charged patch on the surface of the protein (Lys-178–Asp-182–Arg-186) into the buried P450$_{cam}$ active site. This model is supported by recent measurements of solvent isotope effects on

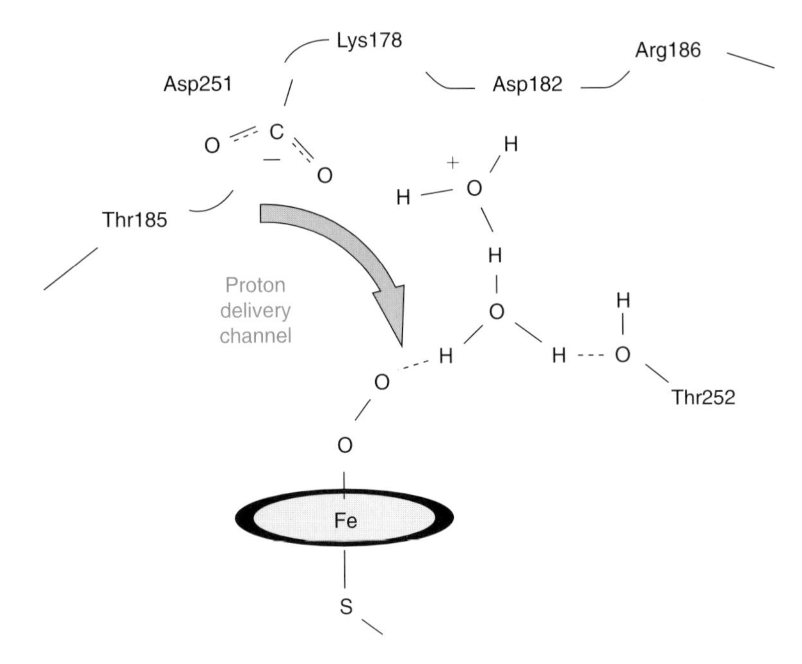

Figure 6. Atmospheric dioxygen activation through acid/base chemistry in the P450 active site

catalysis, the aforementioned mutagenesis studies and detailed X-ray crystallographic investigations by Poulos and Schlichting.

Perspectives

Much remains to be elucidated before a complete picture of cytochrome P450 catalysis is at hand. New reactions of this versatile enzyme continue to be discovered; for example, the ability of this protein to reductively metabolize a variety of halogenated hydrocarbons or quinone-methide pro-drugs that have physiological activity against solid tumours has recently been reported. The identity of the hypothetical intermediates in the reaction cycle illustrated in Figure 5 needs to be confirmed and the detailed electronic configurations defined. Questions of molecular recognition, the specific interactions of redox partners in both membrane and soluble systems, as well as the principles controlling the specificity and stereochemical course of oxygen transfer at the P450 active site still need to be elucidated. One very interesting question concerns how the course of evolution has brought this efficient catalyst to the forefront of metabolic diversity. P450 systems have also been the focus of extensive investigation into both the regulation of transcription and molecular genetics. Finally, numerous research and development directions are evident where this enzymic system is likely to play central roles in biotechnology and pathway engineering for both remediation and chemical synthesis.

Summary

- *Cytochromes P450 are utilized in an enormous diversity of biological reactions, including degradation of xenobiotics, generation of hormones and biosynthesis of a variety of important biological compounds.*
- *The cytochrome P450 family is a major participant in nearly all metabolism of pharmaceutical reagents.*
- *The presence of different P450 enzymes in various quantities in individuals makes the prediction of drug responses in patients highly complex.*
- *A large literature describing mechanistic studies has characterized several intermediates in the oxygenation pathway.*
- *It has recently been shown that two or more possible oxygenated forms of the P450 haem can participate in various oxygenations, with some intermediates being highly electrophilic and others being nucleophilic.*

The work in my laboratory is supported by the National Institutes of Health.

References

1. Ramm, P.J. & Caspi, E. (1969) The stereochemistry of tritium atoms 1, 7 and 15 in cholesterol derived from (3*R*,2*R*)-(2-3H)-mevalonic acid. *J. Biol. Chem.* **244**, 6064–6073
2. Garfinkel, D. (1958) Preparation and properties of a microsomal reduced diphosphopyridine nucleotide – cytochrome *c* reductase. *Arch. Biochem. Biophys.* **77**, 493–509
3. Klingenberg, M. (1958) Pigments of liver microsomes. *Arch. Biochem. Biophys.* **75**, 376–386
4. Hanson, L.K., Eaton, W.A., Sligar, S.G., Gunsalus, I.C., Gouterman, M. & Connell, C.R. (1976) Letter: Origin of the anomalous Soret spectra of carboxycytochrome P-450. *J. Am. Chem. Soc.* **98**, 2672–2674
5. Mason, H.S., North, J.C. & Vanneste, M. (1965) Microsomal mixed-function oxidations: the metabolism of xenobiotics. *Fed. Proc. Fed. Am. Soc. Exp. Biol.* **24**, 1172–1180
6. Champion, P.M., Stallard, B.R., Wagner, G.C. & Gunsalus, I.C. (1982) Resonance Raman detection of an iron–sulphur bond in cytochrome P450cam. *J. Am. Chem. Soc.* **104**, 5469–5472
7. Poulos, T.L., Finzel, B.C. & Howard, A.J. (1987) High-resolution crystal structure of cytochrome P450cam. *J. Mol. Biol.* **195**, 687–700
8. Dawson, J.H., Andersson, L.A. & Sono, M. (1982) Spectroscopic investigations of ferric cytochrome P-450-cam ligand complexes. Identification of the ligand trans to cysteinate in the native enzyme. *J. Biol. Chem.* **257**, 3606–3617
9. Cooper, D.Y., Levin, S., Narasimhulu, S., Rosenthal, O. & Estabrook, R.W. (1965) Photochemical action spectra of the terminal oxidase of mixed function oxidase systems. *Science* **147**, 400–402
10. Lu, A. & Coon, M.J. (1968) Role of hemoprotein P-450 in fatty acid omega-hydroxylation in a soluble enzyme system from liver microsomes. *J. Biol. Chem.* **243**, 1331–1332
11. Hedegaard, J. & Gunsalus, I.C. (1965) Mixed function oxidation. IV. An induced methylene hydroxylase in camphor oxidation. *J. Biol. Chem.* **240**, 4038–4043
12. Olsen, G.T. & Woese, C.R. (1997) Archaeal genomics: an overview. *Cell* **89**, 991–994
13. Gutell, R.R., Larsen, N. & Woese, C.R. (1994) Lessons from an evolving rRNA: 16 S and 23 S rRNA structures from a comparative perspective. *Microbiol. Rev.* **58**, 10–26
14. Shumada, H., Sligar, S.G., Yeom, H. & Ishimura, Y. (1996) Heme monooxygenases. A chemical mechanism for cytochrome P450 oxygen activation, in *Oxygenases and Model Systems* (Funabiki, T., ed.), pp. 195–221, Kluwer Academic Publishers, Hingham, MA
15. Martinis, S.A., Atkins, W.M., Stayton, P.S. & Sligar, S.G. (1989) A conserved residue of cytochrome P-450 is involved in heme-oxygen stability and activation. *J. Am Chem. Soc.* **111**, 9252–9253
16. Groves, J.T. (1985) Key elements of the chemistry of cytochrome P450. The oxygen rebound mechanism. *J. Chem. Educ.* **62**, 928–931
17. Collman, J.P., Chien, A.S., Eberspacher, T.A. & Brauman, J.I. (1998) An agostic alternative to the P-450 rebound mechanism. *J. Am Chem. Soc.* **120**, 425–426
18. Yoshizawa, K. (1998) Two-step concerted mechanism for alkane hydroxylation on the ferryl active site of methane monooxygenase. *J. Inorg. Biol. Chem.* **3**, 318–324
19. Coon, M.J., Vaz, A.D.N., McGinnity, D.F. & Peng, H.-M. (1998) Multiple activated oxygen species in P450 catalysis: contributions to specificity in drug metabolism. *Drug Metab. Disp.* **26**, 1190–1193
20. Newcomb, M., Le Tadic-Biadatti, M.H., Chestney, D.L., Roberts, E.S. & Hollenberg, P.F. (1995) A nonsynchronous concerted mechanism for cytochrome-P450 catalysed hydroxylation. *J. Am. Chem. Soc.* **117**, 12085–12091
21. Ortiz de Montellano, P.R. (ed.) (1995) *Cytochrome P450: Structure, Mechanism and Biochemistry*, 2nd edn., Plenum, New York
22. Gray, H.B. & Winkler, J.R. (1996) Electron transfer in proteins. *Annu. Rev. Biochem.* **65**, 537–561
23. Moser, C.C. & Dutton, P.L. (1996) Outline of theory of protein electron transfer. In *Protein Electron Transfer* (Bendal, D.S., ed.), pp. 1–21, BIOS Scientific Publishers, Oxford

6

Oxygen-carrying proteins: three solutions to a common problem

Donald M. Kurtz, Jr.

Department of Chemistry and Center for Metalloenzyme Studies, University of Georgia, Athens, GA 30602, U.S.A.

Introduction

Both vertebrates and invertebrates derive most of their energy by 'combustion' of organic compounds, and thus require O_2. Given the surface-to-volume ratio of most higher organisms, simple diffusion of O_2 across body surfaces at ambient partial pressures would result in internal O_2 concentrations that are insufficient to sustain life. Furthermore, in response to various stresses, certain tissues or organs require a rapid infusion of O_2, for which passive diffusion would be insufficient. The existence of O_2-carrying proteins and their presence in large concentrations (approaching 20 mM for haemoglobin in red blood cells [1]) in higher organisms is thus rationalized. These proteins can be defined as those capable of undergoing the reversible reaction shown in eqn. (1) to some measurable degree at ambient temperatures and partial pressures of O_2:

$$\text{Protein} + O_2 \rightleftharpoons \text{Protein} - O_2 \quad (1)$$

Three types of O_2-carrying protein are known: haemoglobin (Hb)/myoglobin (Mb), haemerythrin (Hr)/myohaemrythrin (myoHr) and haemocyanin (Hcy). Mammalian species contain Hb/Mb as the only known O_2-carrying proteins. The Hr/myoHr family has so far been found only in marine invertebrate species, and Hcy occurs in arthropods and molluscs. This Chapter attempts to summarize the current state of knowledge as well as some recent advances in our understanding of chemical and biochemical aspects of O_2-carrying proteins, and focuses on the active sites of these proteins. Since all of these proteins contain either iron or copper at their active sites, some relevant chemical properties of O_2 and transition-metal ions are first summarized.

Reactivity of molecular oxygen with transition-metal ions

Thermodynamically, molecular oxygen has the capacity to be quite reactive, but kinetically it is quite sluggish in its reactions with most organic molecules. These two seemingly contradictory properties are rooted in the energies of the highest-occupied molecular orbitals (HOMOs) and lowest-unoccupied molecular orbital (LUMO) of the dioxygen molecule, which are shown in Figure 1. Since most organic molecules have paired electrons, i.e. singlet ground states, they are spin-forbidden from reacting with the triplet ground state of O_2. Complexation of dioxygen with transition-metal ions having unpaired electrons provides a relatively low-energy pathway for overcoming the spin restrictions to reactivity and, thus, the kinetic inertness of O_2. This low-energy pathway derives from overlap of metal-ion d-orbitals with the HOMOs and LUMO on O_2 upon complexation, and this overlap also provides a facile pathway for exchange of electrons between the metal ions and dioxygen, as shown schematically in Figure 2. Given the propensity of O_2 to accept electrons under ambient conditions and the multiple, readily accessible, positive oxidation states of many transition-metal ions, complexation usually results in net transfer of electron density from metal ion to dioxygen. As

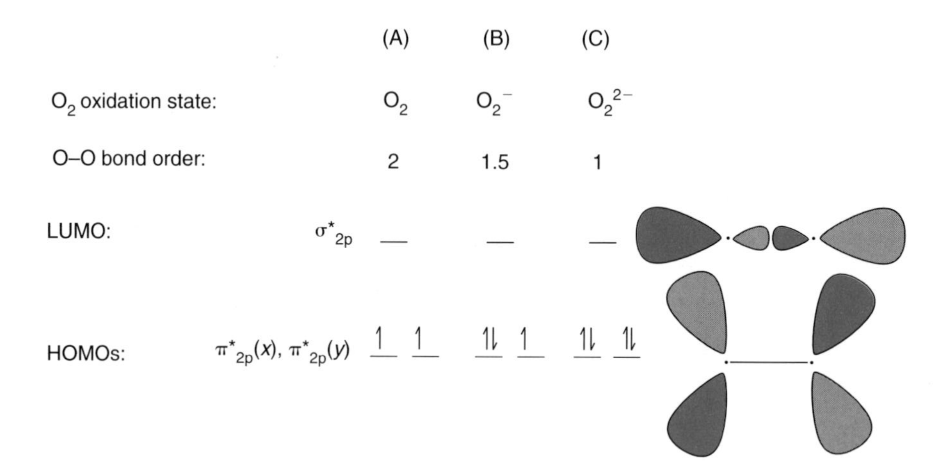

Figure 1. HOMOs and LUMO of the dioxygen molecule and its one- and two-electron reduced forms

(**A**) Dioxygen; (**B**) superoxide; (**C**) peroxide. Shapes of the HOMOs and LUMO are shown to the right of the π^* and σ^* energy levels, respectively, with the O–O bond axis (defined as the z-axis) oriented horizontally. Only one of the two identically shaped HOMOs is shown. The other HOMO would have its lobes oriented perpendicular to the plane of the page.

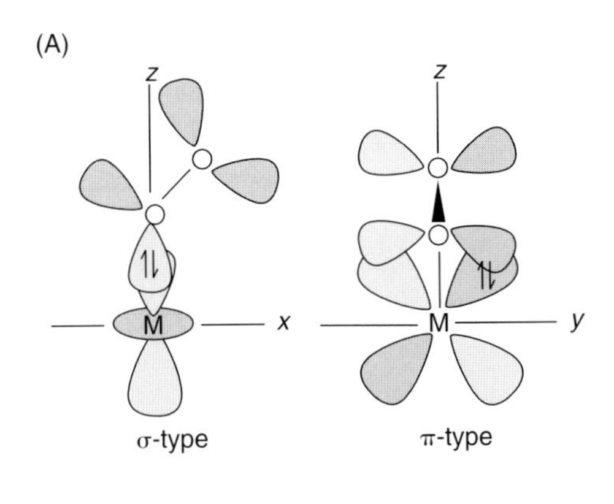

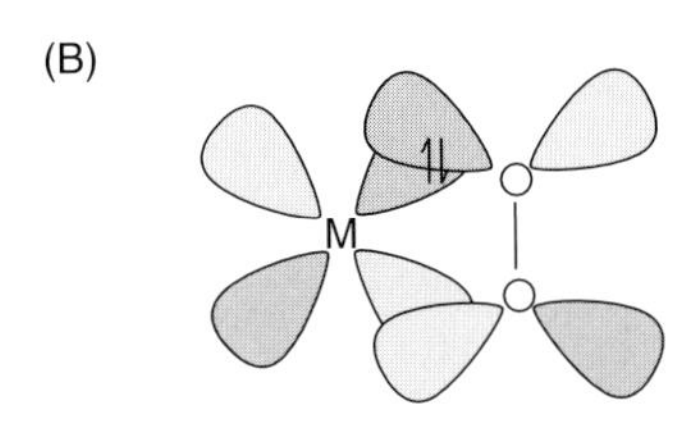

Figure 2. Molecular orbital description of metal–dioxyen adducts
Bonding of O_2 to metal ions (M) in (**A**) end-on, and (**B**) side-on modes is illustrated.

shown in Figures 1 and 2, this electron transfer occurs into effectively antibonding orbitals on O_2. This transfer of electron density manifests itself in lower O–O stretching frequencies and longer O–O distances (due to weaker O–O bonds) in metal–dioxygen adducts compared with molecular oxygen (see Table 1). However, in order for the O_2 binding to be reversible, the extent of electron transfer from metal ion to dioxygen must be delicately balanced, i.e. enough to form a stable complex, but not so much that the O–O bond is cleaved irreversibly.

Table 1. Metal-ion complexation weakens the O–O bond [2]

Species	νO–O (cm^{-1})*	O–O Bond length (Å)
Dioxygen	1555	1.21
Metal–superoxo	1100–1150	1.24–1.31
Metal–peroxo	800–900	1.35–1.5

*O–O Stretching frequency in wavenumbers.

The active sites of oxygen-carrying proteins

Studies of synthetic metal–dioxygen interactions (reviewed extensively in [3]) have shown that oxygen-carrying proteins modulate the delicate balance referred to above in at least three general ways: the choice of metal ion, the type of complex (i.e. type and number of ligands to the metal ion, coordination geometry etc.), and the environment of the complex. The natural selection of iron (in Hb/Mb and Hr/myoHr) and copper (in Hcy) as the metal ions in oxygen-carrying proteins could not necessarily have been predicted from studies of synthetic complexes. Although synthetic iron and copper complexes that reversibly bind O_2 are known, synthetic cobalt complexes that reversibly bind O_2 have also been known for many years, but have no known biological counterpart. One rationale for Nature's selection of iron and copper over other transition metals for reversible O_2 binding is that the large concentrations of O_2-binding proteins necessary for life in higher organisms can be attained only by using metals with relatively high bio-availabilities, such as copper and particularly iron [4].

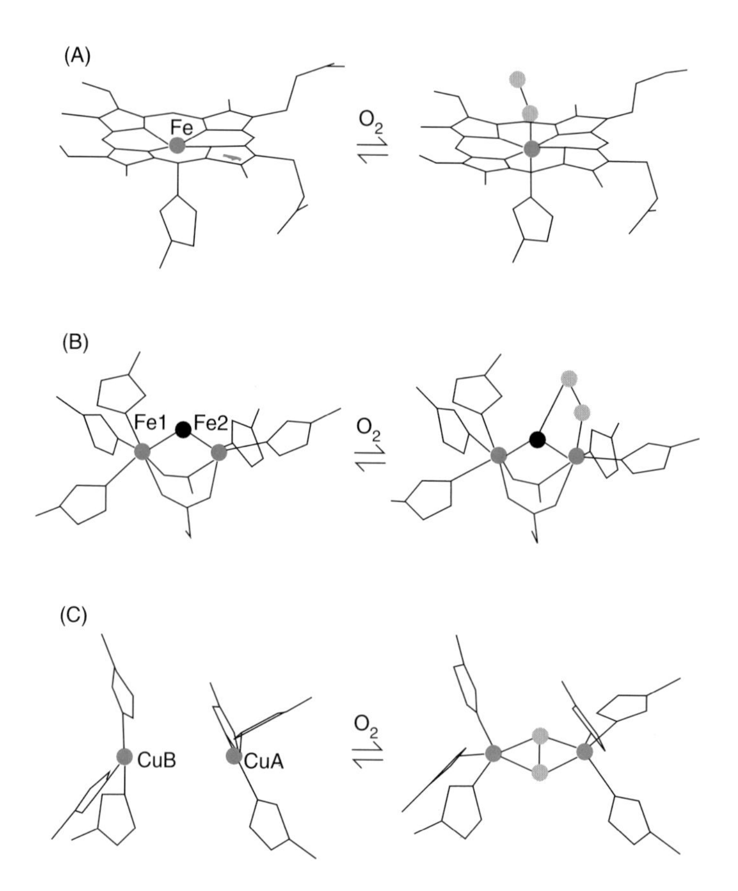

Figure 3. The active site structures of O2-carrying proteins
Active sites are shown schematically for **(A)** sperm whale Mb, **(B)** *Themiste dyscritum* ('peanut' worm) Hr, and **(C)** *Limulus polyphemus* (horseshoe crab) Hcy. Deoxy forms are shown on the left, and oxy forms on the right. Metal ligands are represented as atom-undifferentiated line drawings, and are identified in the text. Blue spheres represent either iron or copper ions, and grey spheres represent bound dioxygen atoms. Black spheres in **(b)** represent oxo or hydroxo ions. The drawings are based on X-ray crystal structure co-ordinates deposited in the Brookhaven Protein Data Bank (Mb, 1MBD, 1MBO [5,6]; Hr, 1HMD, 1HMO [7]; Hcy, 1LLA, 1OXY [8,9]).

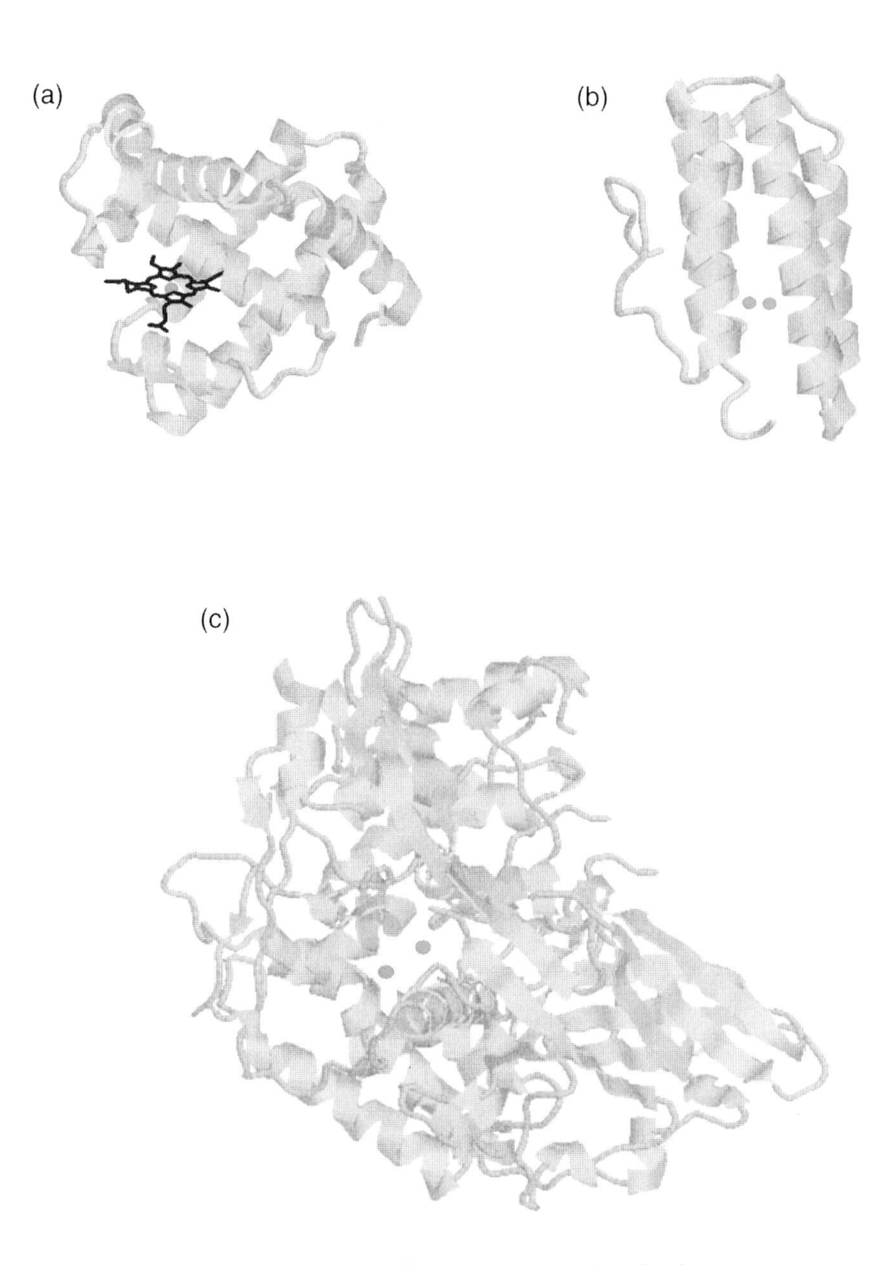

Figure 4. The backbone structures of O2-carrying protein subunits
(**a**) Sperm whale Mb, (**b**) *T. dyscritum* Hr, and (**c**) *L. polyphemus* Hcy. Deoxy forms are shown. Spheres represent either iron (Mb, Hr) or copper (Hcy) ions. The drawings are based on X-ray crystal structure co-ordinates (see Figure 3 legend for references) and are scaled to the relative sizes of the subunits.

Fairly high-resolution X-ray crystal structures are now available for all three types of O_2-carrying protein in both their oxy and deoxy forms. The deoxy and oxy active-site structures are shown in Figure 3, and the backbone structures of the protein subunits are shown in Figure 4. Even a cursory inspection of these structures leads to an obvious conclusion: the three types of oxygen-carrying protein bear no detectable structural similarities to each other, at either subunit-fold or active-site levels. This conclusion is borne out upon detailed inspection as well. Thus although helical regions of the polypeptides surround the active sites in all three types of oxygen-carrying protein, and all of the active sites contain at least one histidine ligand, there is no detectable amino acid sequence homology among them. Clearly, Nature has found three distinct solutions to the problem of binding O_2 reversibly.

Hb and Mb

Hb and Mb incorporate the macrocyclic tetrapyrrole, protoporphyrin IX, which binds a single iron ion via the four pyrrole nitrogen atoms located near its centre. The resulting chelate, called haem, is the prosthetic group that reversibly binds O_2 to the distal side of the haem, as shown in Figure 3. A histidine ligand from the protein (referred to as the 'proximal' histidine) forms a fifth Fe–N bond, which is approximately perpendicular to the haem plane. Hb is tetrameric ($\alpha_2\beta_2$), and each subunit is structurally and functionally very similar to the monomeric Mb. Hb circulates through the body within red blood cells, whereas Mb occurs within muscle tissues. The literature on Hb and Mb is amongst the most voluminous on any protein, and even a relatively limited survey would be impossible in this short essay. Instead, the current status of two intensively studied, but still incompletely resolved, issues about O_2 binding to Hb and Mb will be discussed: (i) the electronic distribution within the Fe–O_2 unit, and (ii) discrimination between O_2 and CO binding.

The electronic distribution within the Fe–O_2 unit

Pauling and Coryell's classic magnetic susceptibility measurements, published in 1936, showed that, whereas deoxyHb is paramagnetic, oxyHb is diamagnetic, consistent with spin pairing of all electrons [10]. This result initiated a controversy about both the electronic distribution and the geometry of the Fe–O_2 unit in Hb. The fascinating history of this controversy, including unsubstantiated challenges to Pauling and Coryell's results, has recently been summarized by Momenteau and Reed [11], and also by Bytheway and Hall [12]. The question of geometry has been settled largely by X-ray crystallography, which clearly shows that O_2 binds to the haem iron in a bent end-on manner in oxyHb, oxyMb and synthetic haem–O_2 adducts. This geometry is illustrated for oxyMb in Figure 3. The reported Fe–O–O angles are within the range of 115–160° [6], which, given the experimental uncertainty, are consistent with the molecular orbital description for end-on O_2 σ-type bonding in these complexes (Figure 2a).

The paramagnetism of deoxyHb can be rationalized readily and quantitatively in terms of the four unpaired d-electrons expected for high-spin Fe(II)

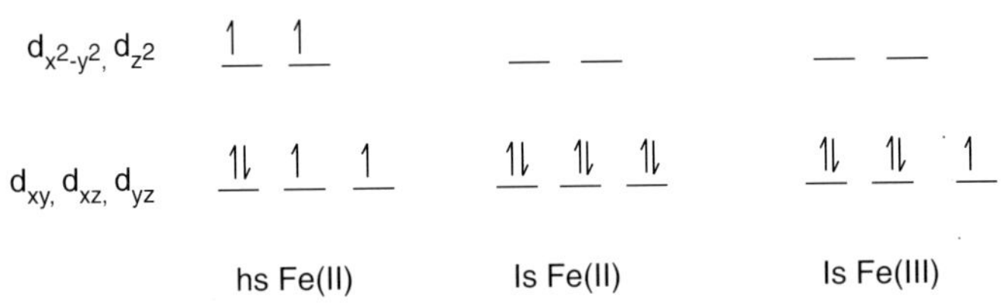

	hs Fe(II)	ls Fe(II)	ls Fe(III)
$d_{x^2-y^2}, d_{z^2}$	$\uparrow$ $\uparrow$	— —	— —
d_{xy}, d_{xz}, d_{yz}	$\uparrow\downarrow$ $\uparrow$ $\uparrow$	$\uparrow\downarrow$ $\uparrow\downarrow$ $\uparrow\downarrow$	$\uparrow\downarrow$ $\uparrow\downarrow$ $\uparrow$

Scheme 1. d-Electron configuration for high-spin (hs) Fe(II), low-spin (ls) Fe(II) and low-spin Fe(III)

(Scheme 1), and a wealth of spectroscopic evidence supports this description. The structural results also show that upon conversion of the deoxy to the oxy forms the iron atom moves closer to the porphyrin plane (see Figure 3). This structural change implies a decrease in the radius of the iron ion upon conversion of the deoxy to the oxy forms, and has been interpreted as indicating a change from high- to low-spin state of the iron, and/or oxidation of Fe(II) in the deoxy forms to a higher formal oxidation state [Fe(III)] in the oxy forms. If the iron remains Fe(II) in the oxy forms, the high-to-low-spin conversion would result in its six d-electrons becoming completely spin paired [ls Fe(II) in Scheme 1]. The pair of electrons in the filled π^* orbital of an effectively spin-singlet O_2 would then have to be donated to the d^{z2} orbital on iron, as shown in Figure 2(a) (σ-type), in order to explain the observed diamagnetism. Back donation of d-electrons on iron into the other π^* orbital of O_2, as shown in Figure 2(a) (π-type), could also occur, leading to $Fe(II)–O_2 \leftrightarrow Fe(IV)–O_2^{2-}$ resonance structures.

Alternatively, if, upon O_2 binding, one electron was transferred from Fe(II) to the co-ordinated O_2, the resulting low-spin Fe(III) would have one unpaired d-electron [ls Fe(III) in Scheme 1]. This d-electron could be spin-paired with the remaining unpaired electron on the co-ordinated O_2^- via the orbital overlaps shown in Figure 2(a) (π-type), thereby achieving diamagnetism. The reported O–O distances in oxyHb and oxyMb range from 1.2–1.3 Å [6], which, when compared with the distances listed in Table 1, are consistent with the superoxo formalism. However, the experimental uncertainties on the O–O distances in the proteins are too large to rule out alternative bonding descriptions. More direct evidence that the superoxo formalism accurately describes the oxidation state of co-ordinated O_2 comes from vibrational spectroscopy, which gives an O–O stretch near 1100 cm^{-1} for both oxyHb and oxyMb [11]. As can be seen from Table 1, this frequency is in the range expected for metal–superoxo complexes, and is well separated from that for either molecular O_2 or O_2^{2-}. Thus considerations of basic bonding and reactivity, as well as experimental results, support the notion that some electron transfer occurs from haem–Fe(II) in deoxyHb and deoxyMb to O_2 upon formation of the oxy adducts. Many bioinorganic chemists regard low-spin $Fe(III)–O_2^-$ as the most accurate and useful representation of the electronic distribution in oxyHb and oxyMb, and there seems to be little interest in revisiting this issue. Nevertheless, as for all formal oxidation-state representations, the electronic distribution and charges represent approximations, not literal truth.

Discrimination between O_2 and CO binding

Synthetic penta-co-ordinate haems in solution favour binding of CO over O_2 by a factor of approximately 100000, whereas Hb and Mb bind CO only about 100-fold more strongly than O_2 [1]. Even the latter lower ratio is not sufficient to prevent the well-known artificial poisoning caused by CO binding to Hb and Mb. However, this toxicity would be even more acute were it not for the discrimination against CO imposed by the surrounding protein. This discrimination is particularly important because it minimizes interference with O_2 binding from CO generated by biological processes such as haem degradation. How is the CO/O_2 affinity ratio reduced in Hb and Mb relative to those of synthetic haems? Until recently, the 'textbook' explanation was that residues lining the O_2-binding pocket in these proteins, particularly the conserved 'distal' histidine (not shown in Figure 3), sterically hindered attainment of the preferred linear Fe–C–O geometry observed in the synthetic haem–CO complexes. On the other hand, as discussed above, the Fe–O–O unit prefers to be bent, so that this steric restriction should be less of a hindrance to O_2 binding. Indeed, X-ray crystal structures of CO adducts of Hb and Mb seem to show a bent, i.e. energetically unfavourable, Fe–C–O unit (Fe–C–O angle of 20–40° measuring the C–O bond axis relative to an axis perpendicular to the average plane of the haem atoms).

However, more recent results are at odds with the distal histidine/steric hindrance explanation for inhibition of CO binding. Both vibrational spectroscopy [13] and polarized infrared spectroscopy [14,15] indicate that the Fe–C–O unit is very close to linear, contrary to the X-ray crystallographic results. Furthermore, site-directed mutagenesis of the distal histidine to residues having either larger or smaller side chains does not have the effects on CO affinity expected for the steric-hindrance explanation [1]. For example, replacement of the distal histidine in Mb with a leucine residue resulted in a 30-fold increase in CO affinity. The effect of the distal leucine replacement on CO affinity could be both hydrophobic and steric in nature, i.e. the isopropylmethyl side chain of leucine is both smaller and less polar than the side chain of histidine. If steric restrictions of the distal side chain dominate CO affinity, then replacement of the leucine with progressively smaller aliphatic side chains, namely those of valine (isopropyl), alanine (methyl) and glycine (hydrogen), should produce larger increases in CO affinity compared with the wild-type than does the leucine replacement. In fact, the opposite occurs; the valine, alanine and glycine mutants show *smaller* increases in CO affinity than does the leucine mutant [1]. These results strongly suggest that steric restrictions imposed by the distal side chain do not control CO affinity in Mb. Since CO/O_2 discrimination also needs to be addressed, it is important to note that the distal histidine-to-leucine mutant of Mb showed a 100-fold *decrease* in O_2 affinity, making the discrimination between CO and O_2 similar to that of synthetic penta-co-ordinate haems.

The following revised explanations for the CO/O_2 discrimination in Hb and Mb are emerging. Pocket polarity rather than steric hindrance is a key factor. This polarity influences ligand binding in two ways. First, the wild-type Mb and Hb crystal structures show a water molecule occupying the O_2-binding pocket in the deoxy forms, but this water is not co-ordinated to the haem iron. The CO adducts show no such 'pocket water'; it is apparently displaced upon CO binding. Thus ligand binding is inhibited by polar pocket residues, which provide a favourable environment for the pocket water compared with the relatively non-polar co-ordinated CO. The relatively polar $Fe^{\delta+}$–O–$O^{\delta-}$ unit, on the other hand, is stabilized by hydrogen bonding to the polar distal histidine, to such an extent that the pocket water is more easily displaced. This explanation is consistent with the relative changes in CO and O_2 affinities of the distal histidine-to-leucine Mb mutant discussed above. Results of time-resolved infrared polarization spectroscopy following photolytic cleavage of the Fe–CO bond suggest another factor that may discriminate against CO binding [15]. CO can be induced to dissociate from the haem by short (ps time-scale) pulses of green polarized laser light. The orientation of the dissociated CO with respect to the haem can then be probed by ultra-short (fs time-scale) polarized pulses of infrared light having energy corresponding to the stretching frequency of the C–O bond. These experiments show that shortly (within a ps) after photolytic dissociation, the C–O bond axis lies parallel to the haem plane in a 'docking site' near the binding pocket, and the CO remains there for a few hundred nanoseconds without rebinding to iron. The protein may thus provide an energetic barrier to the preferred perpendicular orientation required for CO binding to the haem iron, but this barrier does not involve steric hindrance by the distal histidine. However, this issue remains controversial; not all scientists in the field are ready to abandon the distal histidine/steric hindrance explanation.

Hr and myoHr

Of the three types of O_2-carrying protein, the molecular details of O_2 binding to the active site of Hr were, chronologically, the next to be clarified. Hr is most often found as an octamer of essentially identical O_2-binding subunits, and is thought to serve primarily as an O_2-storage reservoir in the marine inverterbrates in which it occurs [7]. MyoHr fulfils a function more closely related to that of Mb and is confined to muscle tissues of the same marine invertebrates. The structure of the myoHr subunit and active site are both very similar to those of Hr (cf. Figures 3 and 4). Therefore, the following discussion applies, with very few exceptions, equally well to both Hr and myoHr. The two iron atoms at the active site are bound directly to protein side chains: five histidines, one aspartate and one glutamate, the latter two carboxylates of which bridge the two irons. An accumulation of spectroscopic evidence [16] had established fairly conclusively, prior to the structural results, that the oxidation state changes shown in eqn. (2) accurately describe O_2 binding in Hr and myoHr (where μ indicates a bridging OH^- or O^{2-}):

$$\underset{\text{deoxy}}{[\text{Fe1}^{\text{II}}(\mu\text{–OH})\text{Fe2}^{\text{II}}]} + O_2 \rightleftharpoons \underset{\text{oxy}}{[\text{Fe1}^{\text{III}}(\mu\text{–O})\text{Fe2}^{\text{III}}O_2H^-]} \quad (2)$$

Thus upon binding, O_2 is formally reduced to the peroxide level by the two Fe(II) in deoxyHr. Perhaps the most direct evidence for the formal oxidation state of the bound O_2 in oxyHr comes once again from vibrational spectroscopy, which shows an O–O stretch at 844 cm^{-1} [16]. This frequency clearly indicates a peroxide (O_2^{2-}) oxidation state, as can be seen by comparison with the values in Table 1. Thus eqn. (2) describes an elegant oxidative addition/proton-transfer reaction, which does not directly involve any amino acid residues. Stenkamp et al. [17] first proposed the active-site structures and mechanism shown in Figure 5 in 1985, and subsequent work has confirmed many of their proposals.

The O_2-binding pocket surrounding Fe2 is hydrophobic, with no nearby water, proton donors or nucleophiles (other than the iron ligands) in either the

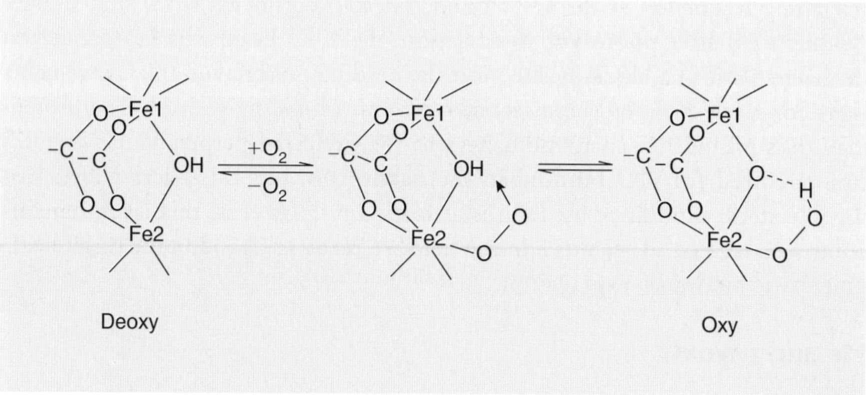

Figure 5. Structural mechanism for O2 binding to the di-iron site of Hr and myoHr
Formal oxidation-state changes of Fe1, Fe2 and O_2 accompany the structural changes, as described in the text. O_2 enters the binding pocket and co-ordinates in a bent end-on fashion to Fe2. This binding initiates oxidation of Fe2, with concomitant shortening of the Fe2–(μ–O) bond (because increasing the metal oxidation state creates a better Lewis acid, thereby attracting nucleophilic ligands more strongly). The increased competition for the electrons on the bridging oxygen causes lengthening and weakening of the O–H bond, and the proton is also attracted to the incipient negative charge developing on the terminal oxygen atom of end-on co-ordinated O_2. The loss of the proton from the bridging oxygen would, in turn, favour shortening of the Fe1–(μ–O) bond, thereby facilitating oxidation of Fe1, even though Fe1 does not directly interact with O_2.

deoxy or oxy forms [7]. Furthermore, the steric constraints of the O_2-binding pocket, together with the five ligands to Fe2 (Figures 3 and 5) in the deoxy form, favour a bent, end-on co-ordination of O_2 to Fe2, but the six ligands to Fe1 and other protein steric constraints greatly inhibit inner-sphere access of O_2 to Fe1. Based on comparisons to synthetic di-iron complexes [18], uninhibited interaction of O_2 with both Fe1 and Fe2 and/or a more polar O_2 pocket would probably make the di-iron site in Hr much more prone to the autoxidation reaction shown in eqn. (3). Autoxidation of Hr does in fact occur, but on the time scale of a day or two at room temperature and ambient partial pressures of O_2. This time scale is several orders of magnitude slower than for the reversible O_2 binding and release reactions represented by eqn. (2).

$$\underset{\text{oxy}}{[\text{Fe1}^{\text{III}}(\mu\text{–O})\text{Fe2}^{\text{III}}\text{O}_2\text{H}^-]} + \text{H}^+ \longrightarrow \underset{\text{met}}{[\text{Fe}^{\text{III}}(\mu\text{–O})\text{Fe}^{\text{III}}]} + \text{H}_2\text{O}_2 \qquad (3)$$

The intermediate depicted between oxy and deoxy in Figure 5 has never been observed, even when using rapid kinetics methods. Thus whether Fe1 is oxidized prior to, concomitant with, or following, transfer of the proton from the oxo bridge to the bound O_2 is not known. Reasonably good evidence exists for formulation of the nitric oxide adduct of deoxyHr as $[\text{Fe1}^{\text{II}}(\mu\text{–OH})\text{Fe2}^{\text{III}}\text{NO}^-]$ [19], suggesting by analogy that $[\text{Fe1}^{\text{II}}(\mu\text{–OH})\text{Fe2}^{\text{III}}\text{O}_2^-]$ is a reasonable formulation for the intermediate, as depicted in Figure 5. However, the failure to detect this (or any other) intermediate implies that the rate-determining step leading to the oxy form occurs either prior to or during formation of this intermediate (with the subsequent conversion to oxy being much faster), and that some combination of reverse proton transfer, electron transfer and cleavage of the Fe–O_2 bond together constitute the rate-determining process for O_2 release. Whereas this rationale is consistent with the available kinetic data [20,21], the elementary steps governing the rates of O_2 binding and release in Hr and myoHr remain to be delineated. Residues lining the O_2-binding pocket, several of which are conserved in all known Hrs and myoHrs, are likely to influence or control some of these steps, and their role is just beginning to be examined [22].

Hcy

Hcys are all large, multidomain, multisubunit proteins and, partly due to their large sizes, the structural and electronic aspects of O_2 binding were the most recent to be clarified among the O_2-carrying proteins. The unusual side-on bridging geometry of O_2 shown in Figure 3 was finally determined by X-ray crystallography in 1994 [8,9]. However, this geometry was accurately predicted from studies of the beautiful model dicopper(II)–peroxo complex synthesized and structurally characterized by Kitajima et al. [23,24]. The diamagnetism and O–O stretching frequency of this synthetic complex also accurately modelled those of oxyHcy and, together with the structure, greatly clarified what had been difficult-to-explain properties of the oxy protein. The 750-cm^{-1} O–O stretching frequency of oxyHcy is unusually low, even for metal–peroxo complexes (see Table 1). Based on the structure of Kitajima's complex, Solomon and co-workers explained this unusually low frequency as due to a π-acceptor interaction of the σ^* orbital on dioxygen (see Figure 1) with d-orbitals on both coppers [25]. This interaction results in transfer of electron density from the two Cu(II) ions to an effectively anti-bonding orbital on O_2^{2-} (i.e. the σ^*), which is normally unoccupied and non-interacting in other metal–peroxo complexes. In terms of weakening the O–O bond strength, this π-acceptor interaction more than compensates for σ *donation* from the effectively π^* orbital on O_2^{2-} to d-orbitals on Cu(II), which occurs for the side-on geometry, as illustrated in Figure 2(b). The long-accepted formal oxidation-state changes embodied in eqn. (4) for the oxygenation reaction of Hcy are, thus, rationalized.

$$[\mathrm{Cu(I), Cu(I)}]+\mathrm{O_2} \rightleftharpoons [\mathrm{Cu(II)}(\mu\text{–}\mathrm{O_2}^{2-})\mathrm{Cu(II)}] \quad (4)$$

Complexes containing a single Cu(II) have one unpaired d-electron; therefore, the observed room-temperature diamagnetism of oxyHcy must be due to spin-pairing of the two originally unpaired d-electrons, one from each Cu(II). Since the CuA–CuB distance of 3.6 Å [8,9] in oxyHcy is considered to be too long for direct overlap of d-orbitals on the two Cu(II), the spin pairing is presumably mediated via relatively weak interactions with paired electrons in orbitals of the bridging peroxide in a phenomenon referred to as superexchange [25]. Whatever the most accurate bonding description may be, it is clear that a (nearly) planar, side-on-O_2-bridged $[\mathrm{Cu(II)}(\mu\text{–}\mathrm{O_2}^{2-})\mathrm{Cu(II)}]$ unit results in unusually extensive orbital overlaps and bonding interactions between metal and dioxygen; nevertheless, this unit can retain reversibility.

X-ray crystal structures of deoxyHcys from two arthropod species both show a dicopper site that appears to be well poised for incorporation of an exogenous bridging ligand (see Figure 3). CuA and CuB are each 3-co-ordinate with no atoms from protein side chains or solvent water visible between the two coppers. In the one case where X-ray crystal structures of the same Hcy (from horseshoe crab) in both forms are available [8,9] (see Figure 3), the CuA–CuB distance decreases by about 1 Å (from 4.6 to 3.6 Å) upon transformation of the deoxy to the oxy form. On the other hand, the side chains of the six histidine ligands move very little upon transformation between oxy and deoxy forms, and the remainder of the tertiary and quaternary structures of the two forms are also very similar to each other. This similarity belies the well-documented co-operativity in O_2 binding exhibited by all Hcys [9]. Co-operativity, in which binding of O_2 to one subunit increases the O_2 affinity of other subunits in the oligomer, is seen to some extent for all multisubunit O_2-carrying proteins, and even a cursory review of this phenomenon would require a separate chapter. A structural mechanism for co-operativity in arthropod Hcys is proposed in [1]. X-ray crystal structures of molluscan Hcys are eagerly awaited, because they could help define a mechanism of co-operativity and also could confirm the presence of an unusual cysteine–histidine thioether bridge to one of the histidine ligands of the dicopper site [26].

Perspectives

How and why did Nature evolve three different solutions to the problem of reversible dioxygen binding? What are the evolutionary antecedents to the three types of O_2-carrying protein? Although fascinating to ponder, these questions may remain forever unanswerable, unless a much more complete picture of evolutionary biology becomes available. A more useful pursuit and realistically attainable goal is probably an understanding of the catalytic chemistry occurring in enzymes with active sites similar to those found in O_2-carrying proteins. Rather than reversibly binding O_2, these enzymes use O_2 as one substrate to oxidize a second substrate, e.g. hydrocarbons, and the enzymic reactions are invariably accompanied by O–O bond cleavage. Parallel examples are known of these so-called 'O_2-activating' enzymes with active sites closely resembling those in each type of O_2-carrying protein (Table 2) [27–30]. These parallel examples suggest that the three types of metal site shown in Figure 3 are particularly well-suited to dealing with the chemistry of dioxygen. Subtle alterations in the environment of the metal ion–dioxygen complex within the O_2-activating enzymes must encourage the thermodynamic propensity of O_2 towards its further reduction and, simultaneously, channel this propensity towards biochemically useful oxidations. The existence of O_2-carrying proteins and analogous O_2-activating enzymes apparently represents Nature's tiptoeing along the edges of the energetic barrier separating reversible O_2 binding from O–O bond cleavage without crossing it. Detailed structure–function comparisons of O_2-binding protein/O_2-activating enzyme pairs may ultimately provide a clearer understanding of the factors required to avoid crossing this barrier while maintaining functionality. For example, it is clear, at least in the cases of the haem and non-haem di-iron active sites, that changes in the type and/or number of metal ligands supplied by the protein constitute one such factor [27,29]. These protein/enzyme pairs are also one reason why the electronic distributions within the metal–O_2 units in O_2-carrying proteins merit detailed

Table 2. O_2-carrying protein/O_2-activating enzyme analogues

Excellent reviews of the O_2-activating enzymes can be found in [27–29].

	Function	
Active site	**O_2-Carrier**	**O_2-Activator**
Haem	Hb	Peroxidases, cytochrome P450
Di-iron	Hr	Ribonucleotide reductase, methane mono-oxygenase
Dicopper	Hcy	Tyrosinase, ascorbate oxidase

and accurate descriptions. Only then can we understand how and why this distribution may differ in the O_2-activating-enzyme counterparts.

Finally, perhaps the most intriguing question of all: are there other types of O_2-carrying protein with active sites different from those shown in Figure 3 awaiting discovery?

Summary

- *Nature has used transition-metal ions with unpaired d-electrons to overcome the kinetic inertness of O_2 and to control its thermodynamic tendency towards reduction.*
- *High-resolution X-ray crystal structures of O_2-carrying proteins show that Nature has devised three distinct solutions to the problem of reversible O_2 binding. The three types can be classified according to their active sites: Hb (haem iron); Hr (non-haem di-iron); and Hcy (dicopper).*
- *The reversible O_2 binding to the three types of active site are formally oxidative additions: Fe(II) to Fe(III)–O_2^- for Hb; [Fe(II),Fe(II)] to [Fe(III),Fe(III)O_2^{2-}] for Hr; and [Cu(I),Cu(I)] to [Cu(II)(μ–O_2^{2-}) Cu(II)] for Hcy. In all cases the O–O bond is weakened, but not cleaved, upon binding.*
- *The 'textbook' explanation for discrimination against CO and O_2 binding to Hb has been revised: steric constraints to the preferred linear Fe–C–O geometry imposed by the 'distal' histidine are no longer thought to play a major role. Instead, recent experimental evidence indicates that the polarity of the binding pocket favours the polar Fe–O–O unit over the relatively non-polar Fe–C–O unit, and that a C–O-binding pocket near the haem also inhibits the preferred linear Fe–C–O geometry.*
- *Reversible O_2 binding to the di-iron site of Hr involves an internal proton transfer as well as electron transfer to O_2, but the elementary steps governing the rates of O_2 binding and release, especially the effects of the surrounding protein, remain to be delineated.*
- *An unusual side-on-bonded O_2 that bridges the two copper ions explains both the unusually low O–O stretching frequency and the diamagnetism of oxyHcy.*
- *O_2-activating-enzyme counterparts exist for each of the three known types of O_2-carrying protein. Detailed comparisons of these protein/enzyme pairs are likely to clarify the factors that tune the delicate balance between reversible O_2 binding and controlled O–O bond cleavage.*

Work in the author's laboratory on O_2-carrying proteins and O_2-activating enzymes has been supported generously for many years by the National Institutes of Health and is currently supported by NIH grant GM40388.

References

1. Springer, B.A., Sligar, S.G., Olson, J.S. & Phillips, Jr., G.N. (1994) Mechanisms of ligand recognition in myoglobin. *Chem. Rev.* **94**, 699–714
2. Vaska, L. (1976) Dioxygen-metal complexes: toward a unified view. *Acc. Chem. Res.* **9**, 175–183
3. Klotz, I.M. & Kurtz, Jr., D.M. (eds.) (1994) Metal-dioxygen complexes. *Chem. Rev.* **94**, 567–856
4. Lippard, S.J. & Berg, J.M. (1994) *Principles of Bioinorganic Chemistry*, p. 103, University Science Books, Mill Valley
5. Phillips, S.E.V. (1980) Structure and refinement of oxymyoglobin at 1.6 Å resolution. *J. Mol. Biol.* **142**, 531–554
6. Shaanan, B. (1983) Structure of human oxyhaemoglobin at 2.1 Å resolution. *J. Mol. Biol.* **171**, 31–59
7. Stenkamp, R.E. (1994) Dioxygen and hemerythrin. *Chem. Rev.* **94**, 715–726
8. Magnus, K.A., Hazes, B., Tonthat, H., Bonaventura, C., Bonaventura, J. & Hol, W.G.J. (1994) Crystallographic analysis of oxygenated and deoxygenated states of arthropod hemocyanin shows unusual differences. *Proteins Struct. Funct. Genet.* **19**, 302–309
9. Magnus, K.A., Ton-That, H. & Carpenter, J.E. (1994) Recent structural work on the oxygen transport protein hemocyanin. *Chem. Rev.* **94**, 727–735
10. Pauling, L. & Coryell, C.D. (1936) The magnetic properties and structure of hemoglobin, oxyhemoglobin and carbonmonoxyhemoglobin. *Proc. Natl. Acad. Sci. U.S.A.* **22**, 210–216
11. Momenteau, M. & Reed, C.A. (1994) Synthetic heme dioxygen complexes. *Chem. Rev.* **94**, 659–698
12. Bytheway, I. & Hall, M.B. (1994) Theoretical calculations of metal–dioxygen complexes. *Chem. Rev.* **94**, 639–658
13. Spiro, T.G. & Kozlowski, P.M. (1997) Will the real FeCO please stand up? *J. Biol. Inorg. Chem.* **2**, 516–520
14. Sage, J.T. (1997) Myoglobin and CO: structure, energetics, and disorder. *J. Biol. Inorg. Chem.* **2**, 537–543
15. Lim, M., Jackson, T.A. & Anfinrud, P.A. (1997) Modulating carbon monoxide binding affinity and kinetics in myoglobin: the roles of the distal histidine and the heme pocket docking site. *J. Biol. Inorg. Chem.* **2**, 531–536
16. Sanders-Loehr, J. (1989) Binuclear Iron Proteins. In *Iron Carriers and Iron Proteins* (Loehr, T.M., ed.), pp. 373–466, VCR, New York
17. Stenkamp, R.E., Sieker, L.C., Jensen, L.H., McCallum, J.D. & Sanders-Loehr, J. (1985) Active site structures of deoxyhemerythrin and oxyhemerythrin. *Proc. Natl. Acad. Sci. U.S.A.* **82**, 713–716
18. Feig, A.L. & Lippard, S.J. (1994) Reactions of non-heme iron(II) centers with dioxygen in biology and chemistry. *Chem. Rev.* **94**, 759–805
19. Nocek, J.M., Kurtz, Jr., D.M., Sage, J.T., Xia, Y.-M., Debrunner, P.G., Shiemke, A.K., Sanders-Loehr, J. & Loehr, T.M. (1988) Nitric oxide adducts of the binuclear iron center of hemerythrin. Spectroscopy and reactivity. *Biochemistry* **27**, 1014–1024
20. Projahn, H.-D., Schindler, S., van Eldik, R., Fortier, D.G., Andrew, C.R. & Sykes, A.G. (1995) Formation and deoxygenation kinetics of oxyhemerythrin and oxyhemocyanin. A pressure dependence study. *Inorg. Chem.* **34**, 5935–5941
21. Lloyd, C.R., Eyring, E.M. & Ellis, Jr., W.E. (1995) Uptake and release of O_2 by myohemerythrin. Evidence for different rate-determining steps and a caveat. *J. Am. Chem. Soc.* **117**, 11993–11994
22. Raner, G.M., Martins, L.J. & Ellis, Jr., W.R. (1997) Functional role of leucine-103 in myohemerythrin. *Biochemistry* **36**, 7037–7043

23. Kitajima, N., Fujisawa, K. & Moro-oka, Y. (1989) μ-η^2:η^2-Peroxo binuclear copper complex, $[Cu(HB(3,5\text{-}iPr_2pz)_3)_2O_2]$ *J. Am. Chem. Soc.* **111**, 8975–8976
24. Kitajima, N. & Moro-oka, Y. (1994) Copper-dioxygen complexes. Inorganic and bioinorganic perspectives. *Chem. Rev.* **94**, 737–767
25. Solomon, E.I., Tuczek, F., Root, D.E. & Brown, C.A. (1994) Spectroscopy of binuclear dioxygen complexes. *Chem. Rev.* **94**, 827–856
26. Gielens, C., De Geest, N., Xin, X.Q., Devreese, B., Van Beeumen, J. & Preaux, G. (1997) Evidence for a cysteine-histidine thioether bridge in functional molluscan haemocyanins and location of the disulfide bridges in functional units d and g of the betaC-haemocyanin of *Helix pomatia*. *Eur. J. Biochem.* **248**, 879–888
27. Kurtz, Jr., D.M. (1997) Structural similarity and functional diversity in diiron-oxo proteins. *J. Biol. Inorg. Chem.* **2**, 159–167
28. Solomon, E.I., Sundaram, U.M. & Machonkin, T.E. (1996) Multicopper oxidases and oxygenases. *Chem. Rev.* **96**, 2563–2605
29. Sono, M., Roach, M.P., Coulter, E.D. & Dawson, J.H. (1996) Heme-containing oxygenases. *Chem. Rev.* **96**, 2841–2887
30. Waller, B.J. & Lipscomb, J.D. (1996) Dioxyen activation by enzymes containing binuclear non-heme iron clusters. *Chem. Rev.* **96**, 2625–2657

7

Biological electron-transfer reactions

A. Grant Mauk

Department of Biochemistry and Molecular Biology, University of British Columbia, Vancouver, British Columbia V6T 1Z3, Canada

Introduction

Electron-transfer reactions are characteristic features of a variety of fundamental biological processes that include energy metabolism (photosynthesis, respiration, nitrogen fixation), hormone (steroids, prostaglandins) biosynthesis and xenobiotic detoxification. For most of the proteins involved in these processes, the active site is comprised of a metal centre, although organic cofactors such as flavins or quinones may also fulfil this function. Whereas the mechanistic complexity of biological electron-transfer reactions varies considerably from case to case, the underlying principles that dictate the rate of electron transfer are the same.

The intense research activity that has been directed towards understanding biological electron-transfer processes in recent years reflects the importance of the metabolic processes in which electron transfer is involved and the successful interaction of theoretical methods and innovative experimental strategies to understand them. Perhaps no other area of biochemical research has enjoyed as productive an interaction between theoretical and experimental methods. To understand the basis for this situation and to define many of the mechanistic issues related to biological electron-transfer reactions, it is useful to consider the inorganic origins of current biological research activities.

Inorganic origins

As reviewed by Marcus [1], the genesis of contemporary perspectives of biological electron-transfer reactions can be traced to the late 1940s when co-ordination compounds with radiolabelled transition metals were used initially to study inorganic electron-transfer reactions. With radiolabelled transition-metal complexes, it was possible for the first time to determine rate constants for the transfer of an electron from the reduced form of such a complex to the oxidized form. Previously, the chemical identity of the reactants and the products of such reactions and the limitations of experimental methods of the time had prevented the determination of such rate constants, i.e. electron-transfer self-exchange rate constants. The availability of reliable experimental information for self-exchange reactions combined with the chemical simplicity of electron-transfer reactions (no bonds are created or destroyed and the free energy of the products is identical to that of the reactants) attracted the attention of both experimentalists and theoreticians. Because much of our current understanding of the mechanisms by which electron-transfer proteins function is based on insights gained from the study of inorganic electron-transfer reactions, it is useful to consider briefly some highlights from this early work before considering the related but more complex biological systems.

The experimental work of Taube, Halpern and others (summarized in [2,3]) led to the recognition of two general types of inorganic electron-transfer mechanism. The simpler mechanism is referred to as outer-sphere electron transfer, and it involves three steps:

$$\mathrm{A+D} \rightleftharpoons \mathrm{A\|D} \tag{1}$$

$$\mathrm{A\|D} \xrightarrow{k_{\mathrm{et}}} \mathrm{A^{-}\|D^{+}} \tag{2}$$

$$\mathrm{A^{-}\|D^{+}} \rightleftharpoons \mathrm{A^{-}+D^{+}} \tag{3}$$

The first of these steps (eqn. 1) is the formation of the so-called precursor complex (A||D), in which the electron acceptor (A) and electron donor (D) interact through the ligands co-ordinated to the central metal atom. Following formation of the precursor complex, the electron is transferred from the electron donor to the electron acceptor, with the rate constant k_{et}, to form the successor complex ($A^{-}\|D^{+}$). The reaction is completed with the dissociation of the successor complex to the products of the reaction. Whereas the rate of intramolecular electron transfer will be influenced only by factors that affect k_{et}, the intermolecular reaction will also be a function of factors that affect the interaction of the reactants with each other (diffusion of the reactants towards each other for formation of the precursor complex and electrostatic interactions between the reactants).

On the other hand, transient formation of a reaction intermediate in which the inner co-ordination spheres of the electron donor and electron acceptor metal ions share a common ligand is characteristic of inner-sphere electron-transfer reactions. One example of a generalized inner-sphere mechanism is illustrated below:

$$Y_nA\text{–}X + Y'_mD(H_2O) \rightleftharpoons [Y_nA\text{–}X\text{–}DY'_m] + H_2O \quad (4)$$

$$[Y_nA\text{–}X\text{–}DY'_m] \xrightarrow{k_{et}} [Y_nA^-\text{–}X\text{–}D^+Y'_m] \quad (5)$$

$$[Y_nA^-\text{–}X\text{–}D^+Y'_m] \rightleftharpoons Y_nA^-(H_2O) + X\text{–}D^+Y'_m \quad (6)$$

As before, A is the electron acceptor, and D is the electron donor. In this particular example, the bridging ligand X is initially associated with the electron acceptor, and this ligand is transferred in concert with the electron. Whereas an inner-sphere electron-transfer mechanism can be assigned unequivocally to such reactions, transfer of the bridging ligand is not a requirement for an inner-sphere process. Furthermore, the bridging ligand could equally well be associated initially with the electron donor.

As might be expected from these two types of mechanism, co-ordination complexes that are relatively inert to ligand substitution employ outer-sphere electron-transfer mechanisms, and those complexes in which co-ordinated ligands are more labile are more prone to undergoing inner-sphere electron-transfer reactions. The requirement for formation of the ligand-bridged intermediate prior to electron transfer means that the energetic barrier to formation of the transition state for the inner-sphere reaction is significantly greater than that required for formation of the transition state for outer-sphere electron transfer. This difference in activation barriers between the two mechanisms is a manifestation of the Franck–Condon principle, which acknowledges that nuclear rearrangement is slower than electronic rearrangement.

Theoretical basics

The relative mechanistic simplicity of outer-sphere electron-transfer reactions provided an attractive experimental basis for development of theoretical efforts towards the prediction of electron-transfer rate constants. Several descriptions of varying depth concerning the relevant theory have been published previously (e.g. [1,4–7]), so only the basics will be considered here. A convenient starting point for the current description is provided by considering the description of reactions such as eqn. (2) by transition-state theory, as depicted in Figure 1. In this diagram, the potential energies of the precursor complex (A||D) and the successor complex (A^-||D^+) are depicted in two dimensions as a function of an ill-defined nuclear co-ordinate.

Proximity of the acceptor and donor centres to each other within the A||D complex results in electronic interaction between the centres that produces

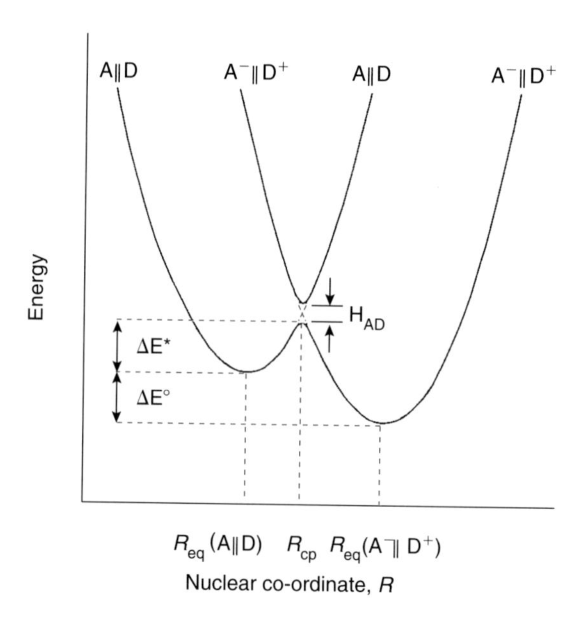

Figure 1. Potential energy diagram for an electron-transfer reaction
The parabola on the left represents the potential energy of the reactants (A||D), and the parabola on the right represents the potential energy of the products ($A^-||D^+$). The activation energy for the electron-transfer reaction is indicated by ΔE^*, the thermodynamic driving force for the reaction is indicated as ΔE°, and the electronic interaction between the electron donor and acceptor centres is indicated as H_{AD}.

mixing and splitting of the two potential energy surfaces in the region where they intersect (the crossing point, R_{cp}). The greater this electronic interaction or coupling, the greater the separation between the upper and lower curves (ΔE°) and the greater the probability that the reactants will proceed from the A||D curve to the $A^-||D^+$ curve at R_{cp}. Reactions characterized by strong electronic interaction or coupling between the acceptor and donor centres are referred to as adiabatic because the probability that the activated complex will proceed to products is, or is near, unity. For A||D complexes in which electronic coupling between the centres is poor, as will occur when the distances separating them are sufficiently great, the separation between the upper and lower potential energy curves of Figure 1 is small, and there is an increased probability that the reactants will not progress to the $A^-||D^+$ curve at the crossing point. Reactions of this type are referred to as non-adiabatic. Compared with the metal centres in co-ordination compounds, the environments of metal ions bound to active sites of metalloproteins are relatively insulating in character, and the metal ions are generally separated from each other by relatively large distances. As a result, transfer of an electron from one such centre to another is relatively non-adiabatic in nature. Mechanistic analysis of such reactions is usually discussed in terms of electron tunnelling.

Correlation of experimental electron-transfer rate constants for non-adiabatic reactions with those calculated by theory requires consideration of: (i) the magnitude of the electronic interaction between the donor and acceptor centres as described by the electronic coupling matrix element H_{DA} (the separation between the A||D and A^-||D^+ curves at the crossing point in Figure 1); and (ii) the contribution of the Franck–Condon factor (*FC*). These considerations are usually accounted for through use of Fermi's golden rule:

$$k_{et}=(2\pi/\hbar)\ |H_{DA}|^2 FC \qquad (7)$$

The magnitude of the matrix element H_{DA} is dependent on both the distance between, and the nature of, the donor and acceptor centres and can be described by the relationship of Gamow:

$$|H_{DA}|^2=|H_M|^2\cdot\exp(-\beta R) \qquad (8)$$

where R is the distance separating the electron donor and acceptor centres, β is the exponential coefficient that describes the decay of electronic coupling with R, and H_M is the value of the matrix element for maximum electronic coupling. The Franck–Condon term, *FC* in eqn. 7, depends on: (i) the energy λ required for the rearrangement of the atomic nuclei of the reactants into the configuration that they occupy in the products; and (ii) the equilibrium constant or thermodynamic driving force for the reaction, $-\Delta G^\circ$, which is derived from the difference in midpoint reduction potentials of the electron donor and acceptor centres. Marcus developed the classical expression for *FC* shown below:

$$FC=(4\pi\lambda k_B T)^{-1/2}\cdot\exp[-(\lambda+\Delta G^\circ)^2/4\lambda k_B T] \qquad (9)$$

in which λ is the reorganization energy, k_B is Boltzmann's constant and $-\Delta G^\circ$ is the thermodynamic driving force of the reaction. Alternative, quantum-corrected, functions for *FC* have also been reported as reviewed elsewhere [4–7]. For both the classical and quantum-mechanical treatments, it can be seen that ln k_{et} varies parabolically with $-\Delta G^\circ$, such that k_{et} increases with the driving force of the reaction until $-\Delta G^\circ=\lambda$. As the driving force increases beyond this point, k_{et} decreases (providing λ remains constant). This remarkable prediction, that the rate constant eventually decreases with increased driving force, constitutes what is referred to as the Marcus inverted region.

From these considerations, the two dominant factors that contribute to the rate of electron transfer are the distance separating the donor and acceptor centres and the thermodynamic driving force of the reaction. The critical nature of these dependencies in the assessment of various theoretical considerations and the behaviour of biological electron-transfer proteins has stimulated considerable experimental effort to quantify the nature of these relationships and to verify the existence of the inverted region by experiment. For more detailed treatments of these and other critical theoretical issues, specialized reviews should be consulted [1,4–7].

One consequence of the considerations presented above and first noted by Marcus is that the cross-reaction rate constant (k_{12}) for electron transfer between two substitutionally inert complexes (i.e. the reorganization term λ is nearly constant) is a function of the self-exchange rate constants for each of the reactants (k_{11} and k_{22}) and the difference in reduction potentials of the reactants (i.e. the equilibrium constant of the reaction, K_{12}, which is often referred to as the thermodynamic driving force of the reaction):

$$k_{12}=(k_{11}k_{22}K_{12}f)^{1/2} \quad (10)$$

where

$$\log f=(\log K_{12})^2/4\log(k_1k_2/z^2) \quad (11)$$

and z is the frequency of collision between electrostatically neutral molecules in aqueous solution and is often estimated to be $\approx 10^{11}\ M^{-1}\cdot s^{-1}$. Several authors have subsequently reported alternative derivations of this relationship, and a particularly clear account has been provided by Newton [8]. The simplicity of 'relative' Marcus theory, as described by these equations, rendered it amenable to experimental evaluation, in part because the self-exchange rate constants for a variety of inorganic complexes were reported during the 1950s and early 1960s. With these values, it was possible to predict rate constants (within an order of magnitude) for 'cross-reactions' between two co-ordination complexes for which self-exchange rate constants and reduction potentials were known. These initial experimental tests of relative Marcus theory were provided by stopped-flow-kinetics studies of Sutin [9]. These results and those provided by other laboratories over several years of work provided compelling evidence for the validity of this relationship for reactions involving inorganic complexes.

Electrochemistry and biological electron-transfer kinetics

From the synopsis of electron-transfer theory provided above, it is apparent that one of the major factors that contributes to an electron-transfer rate constant is the equilibrium constant (the thermodynamic driving force) for the reaction. As a result, any rigorous analysis of electron-transfer kinetics requires knowledge of the equilibrium constants for the reactions under study. The standard free energy of an oxidation–reduction reaction is defined by the difference between the reduction potentials of the two reactants [$\Delta G^\circ=-nF(\Delta E)=-23.06(\Delta E)$ kcal/mol (T=298 K), where n is the number of electrons transferred, F is the Faraday constant and ΔE is expressed in volts, an electron volt being the equivalent of 23.06 kcal/mol, the energy acquired by an electron while passing through a potential of 1 V]. Therefore, the equilibrium constant for the reaction can be calculated in a straightforward fashion

($\Delta G^\circ = -RT\ln K$) if the reduction potentials of the reactants are known. Since the pioneering studies of J.B. Conant concerning the oxidation–reduction equilibrium of haemoglobin in the 1920s, several methods have been used to study the electrochemical properties of proteins that possess metals, flavins and other oxidation–reduction centres at their active sites. At present, two methods are used most frequently, and these methods in some respects provide complementary types of information.

Spectroelectrochemistry provides a convenient means of performing potentiometric titrations while monitoring the electronic absorption spectrum of the protein [10,11]. The principle underlying spectroelectrochemical titrations is identical to that of more traditional methods in which the potential of the protein solution is monitored as the solution potential is titrated with a reducing agent or an oxidizing agent. With spectroelectrochemical methods, however, the oxidation state of the protein can be monitored spectrophotometrically and, as will be seen, the titrant (usually a relatively complex organic compound) is not required. Because spectroelectrochemical methods monitor the response of the oxidation–reduction equilibrium of the protein to incremental changes in the potential of the solution, this technique is sometimes referred to as an 'equilibrium' method.

The spectroelectrochemical approach most commonly involves use of a mini-grid of closely spaced gold wires (≈120–200/cm) as a 'working' electrode that allows transmission of ≈60% of incident light. This semi-transparent electrode is placed between two quartz windows to create what amounts to a small cuvette. The design of the resulting semi-transparent optical cell also accommodates a reference electrode (usually calomel or Ag/AgCl) and a counter electrode (usually platinum). The solution potential is controlled by connecting the working electrode to a stable power supply (a potentiostat), the output of which can be adjusted (titrated) to poise the oxidation–reduction equilibrium of the protein solution as desired. The solution potential can be measured with a microvoltmeter in conjunction with the reference electrode. A representative cell of this type [11] is depicted in Figure 2. In the design of this and many other such cells, the quartz windows are separated by a short distance (≈0.02 cm) to ensure that diffusion is not a factor. Cells with such short pathlengths are frequently referred to as optically transparent thin-layer electrodes (OTTLEs). Nevertheless, if the optical properties of the protein necessitate the use of longer optical pathlengths, the design of the semi-transparent electrochemical cell can be modified suitably, but in this case a stirring mechanism is required to ensure equilibration. Other specialized modifications have also been reported to permit data acquisition under a variety of extreme solution conditions.

One challenge to the successful implementation of spectroelectrochemical methods in the study of proteins is the difficulty that proteins exhibit in exchanging electrons effectively with electrode surfaces. In general, proteins will not exchange electrons effectively with gold or other metallic electrode

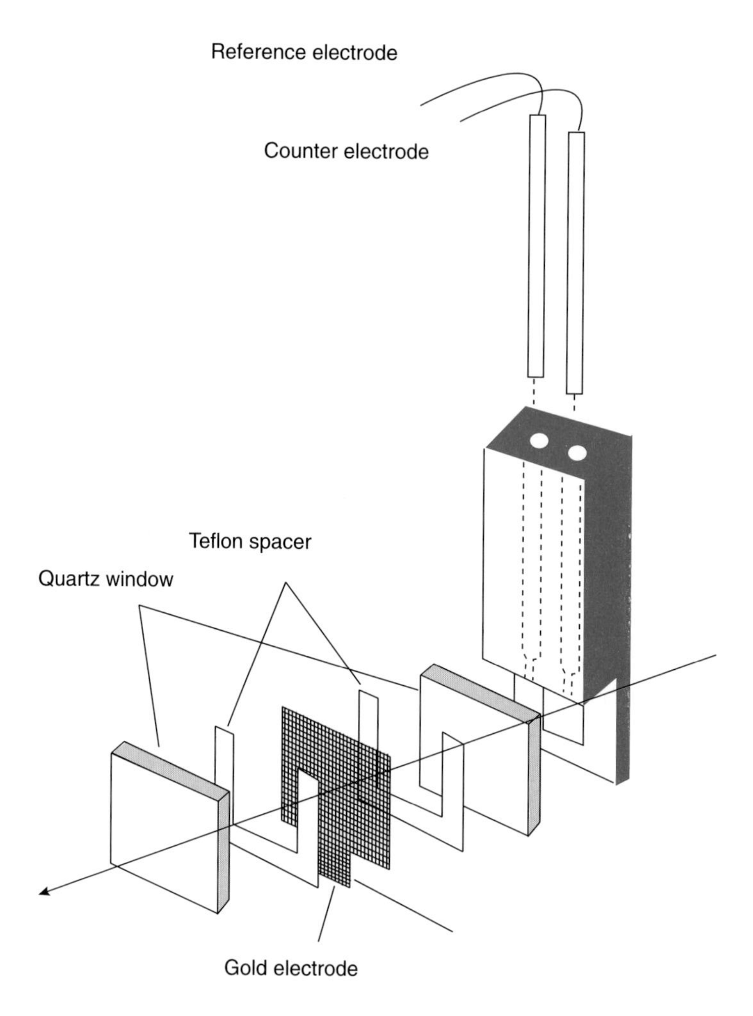

Figure 2. An exploded diagram of a representative optically transparent thin-layer electrode
The gold mini-grid is placed between the quartz windows, and the assembly is mounted on a plexiglass block machined to hold the reference and counter electrodes in contact with the protein solution. The Teflon spacers help define the optical pathlength and contain the protein solution within the sample compartment.

surfaces. To overcome this problem, soluble inorganic or organic compounds that can equilibrate efficiently with such surfaces are employed as 'mediators' that couple the oxidation–reduction equilibrium of the protein to that of the electrode (Scheme 1). Several factors should be considered in selection of a mediator for spectroelectrochemical measurements: (i) the mediator should provide little or no contribution to the electronic spectrum of the solution in the wavelength range of interest; (ii) the mediator should not interact with or bind to the protein in either oxidation state such that it influences the apparent midpoint potential; and (iii) the midpoint potential of the mediator should be

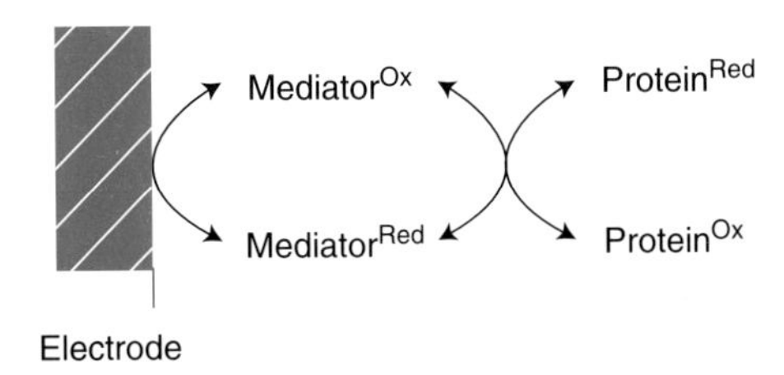

Scheme 1. Mediators facilitate equilibration of electron-transfer proteins with electrode surfaces

similar to that of the protein to ensure efficient equilibration at all potentials required during the potentiometric titration. At times, mixtures of mediators may be required if a single mediator with the optimal midpoint potential is not available. Inclusion of a mediator with a low potential may also be helpful as a means of 'scrubbing' residual oxygen from the system if the presence of oxygen is a problem. Ideally, the concentration of the mediator should be about 10% of the protein concentration used for the spectroelectrochemical titration; however, this factor is not critical as long as higher concentrations do not produce optical interference or increase undesired binding interactions with the protein.

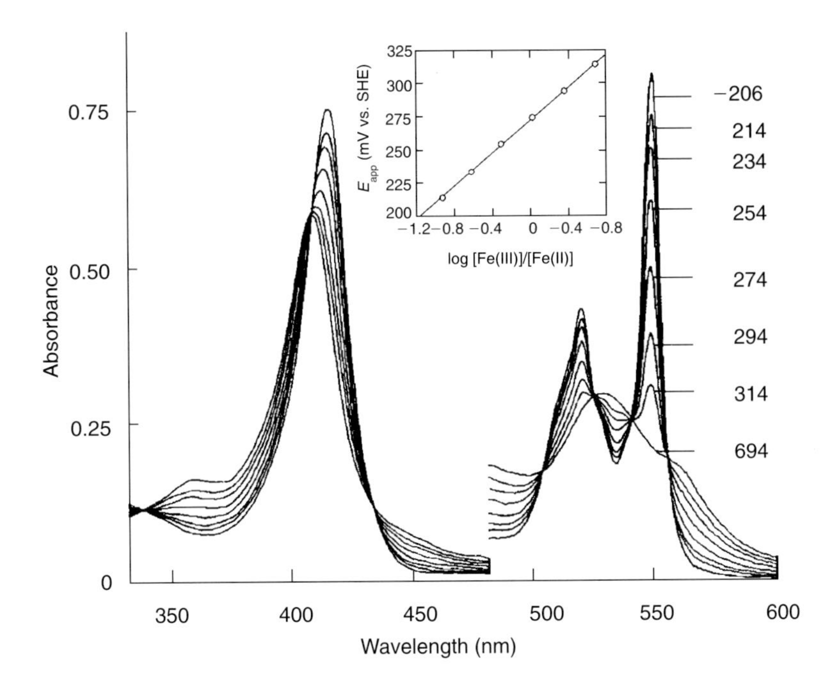

Figure 3. Representative spectroelectrochemical titration of cytochrome c
The solution potential (versus standard hydrogen electrode, SHE) at which each spectrum was recorded is indicated. The Nernst plot for this experiment is shown in the inset.

The results of a representative spectroelectrochemical titration are shown in Figure 3. From such data, it is possible to calculate the relative amounts of reduced and oxidized protein as a function of solution potential. Fitting these data to the Nernst equation:

$$E_{app} = E_m + \frac{2 \cdot 3\,RT}{nF} \cdot log \frac{[Ox]}{[Red]} \qquad (12)$$

permits precise determination of the midpoint potential, E_m, by construction of a Nernst plot (Figure 3, inset) from knowledge of the [Ox]/[Red] ratio at each applied solution potential (E_{app}). The midpoint potential is the solution potential at which the reduced and oxidized forms of the protein are present in equal amounts (i.e. where log[Ox]/[Red]=0). In addition, the slope of the Nernst plot provides an indication of the quality of the data. For most metalloproteins, the number of electrons involved in the oxidation–reduction equilibrium (n in the Nernst equation) is 1, so a slope of 59 mV ($=2.3RT/F$ at 25°C) is indicative of a well-behaved equilibrium and increases confidence in the value obtained for E_m.

The development of direct electrochemical methods in the late 1970s [12,13] provided an alternative to the equilibrium approach represented by spectroelectrochemistry. A typical electrochemical cell used for direct electrochemistry of electron-transfer proteins is illustrated in Figure 4. The success of direct electrochemical methods resulted from development of two alternative methods for overcoming the difficulty that proteins experience in effective interaction with electrode surfaces. One method has been to identify electrode materials with surfaces with which proteins can interact effectively. Surfaces commonly used in such work are tin-doped indium oxide film deposited on glass [12], glassy carbon [14] or edge-oriented pyrolytic graphite [15]. An alternative approach is to use metallic electrode surfaces following chemical treatment. Typically, the surface properties of the electrode, usually gold, are modified by reaction with reagents referred to as 'modifiers'. Those modifiers that result in the preparation of modified electrode surfaces capable of efficient reactions with proteins are referred to as 'promoters', because they promote electrochemistry at the electrode surface. A fundamental distinction between promoters and mediators (used in potentiometric titrations) is that promoters are not electrochemically active. Instead, promoters produce an electrode surface that is conducive to formation of protein–electrode interactions that lead to electron-transfer reactions. Effective promoters are widely divergent in structure, and considerable effort has been directed towards correlating the structures of promoters and properties of proteins the direct electrochemistry of which they promote [16]. However, the identification of an effective promoter to use with a new electron-transfer protein or enzyme remains largely an empirical exercise that is frequently the principal obstacle to the successful implementation of this method.

Data acquired through direct electrochemical measurements are fundamentally different from those obtained by equilibrium methods. The direct

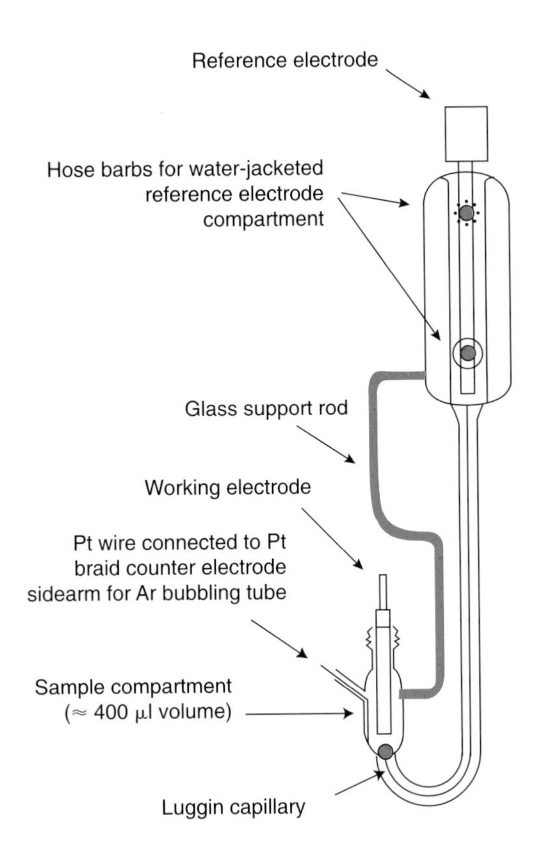

Figure 4. Representative design of an electrochemical cell used for direct electrochemistry

The reference electrode is held at constant temperature in a jacketed enclosure, and the sample compartment is placed into a temperature-controlled water bath without immersing the electrical contacts. The sample is purged gently with argon prior to data collection.

electrochemical experiment records the current passed to or from the solution in response to a change in electrode potential in the form of a cyclic voltammogram (Figure 5). Whereas the theoretical implications of this type of experiment are beyond the scope of the current review (for further discussion see [17,18]), the information derived from a cyclic voltammogram can be considered briefly. As indicated in Figure 5, current is passed as the potential of the solution reaches a value that leads to oxidation or reduction of the electroactive species (the protein) in solution. As the solution potential is decreased, current passes as the protein is reduced (the reduction wave), and as the solution potential is increased, current passes as the protein is oxidized (the oxidation wave). When the minimum of the reduction wave and the maximum of the oxidation wave are separated by 59 mV, then the electrochemical equilibrium being monitored is said to be electrochemically well behaved because it obeys the Nernst equation. For systems that exhibit this type of behaviour, the midpoint potential can be obtained by determining the voltage mid-way

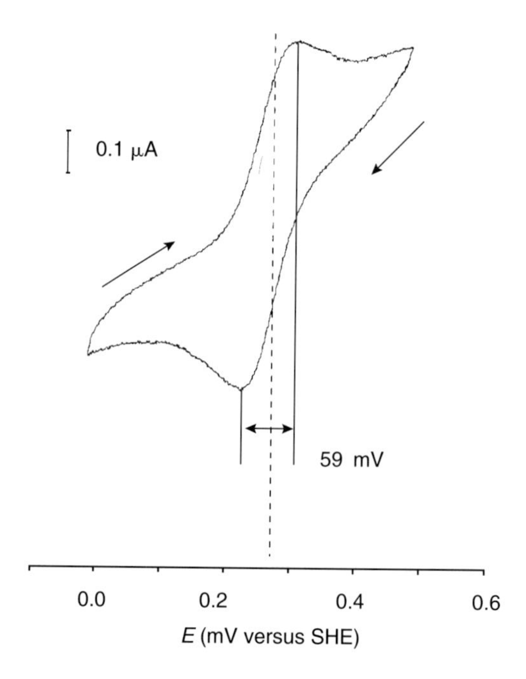

Figure 5. Representative cyclic voltammogram of cytochrome c
Direct electrochemistry was achieved at a gold working electrode modified with 4,4′-dithiodipyridine (20 mV/s). The solution potential was initially positive and was swept to lower values to reduce the protein. When the potential was lowered to ≈0 mV, the direction of the potential sweep was reversed to higher values to re-oxidize the protein. The midpoint potential determined from the minimum and maximum of the current-potential plot is indicated with a dashed line. The peak-to-peak separation of 59 mV (illustrated) was consistent with Nernstian behaviour. SHE, standard hydrogen electrode.

between the minimum observed for the reduction wave and the maximum observed for the oxidation wave. Once a modified electrode surface has been identified that is effective for the protein of interest, data acquisition by this method is far more rapid (a few minutes) than by spectroelectrochemistry (a few hours).

Direct electrochemistry is a dynamic method that permits detection of an electrochemical response of the protein in real time rather than at equilibrium. Whereas the midpoint potentials of electrochemically well-behaved oxidation–reduction equilibria can be determined with greater precision by spectroelectrochemical titrations, direct electrochemistry can provide more mechanistic information because this method permits detection of transiently stable electrochemically active species. Such transient electroactive species may escape detection and may result in behaviour that appears to be non-Nernstian when studied at equilibrium. Examples of such intermediates have been reported for cytochrome *c* at alkaline pH [15] and for a number of other proteins, as reviewed by Armstrong [19,20]. It is likely that as these methods are more widely applied to electron-transfer proteins, direct electrochemistry will become a standard means of obtaining important mechanistic insight into the function of complex oxidation–reduction enzymes.

Early biological kinetic experiments

Electron-transfer proteins in which the active sites are metal centres have many features in common with simple co-ordination complexes. Examples of the co-ordination environments of representative haem iron, copper and iron–sulphur centres found in such proteins are shown in Figure 6 and in many of the Chapters in this volume. Metalloproteins that function solely in electron-transfer reactions possess metal centres that exhibit minimal change in structure with a change in oxidation state. That is, the structures of the reduced and oxidized forms of such proteins are remarkably similar to each other. On the other hand, metalloproteins that undergo electron-transfer reactions as part of a catalytic cycle may exhibit more extensive structural rearrangement in response to a change in oxidation state (e.g. see Chapter 3 in this volume).

One of the first ways in which Marcus theory was applied to the investigation of electron-transfer kinetics of metalloproteins was through analysis of the kinetics by which such proteins react with substitutionally inert co-ordination compounds. Through the initial efforts of Sutin, Gray and others (reviewed by Bennett [21]), the kinetics of a variety of metalloprotein–small-reagent pairs were studied by stopped-flow spectroscopy, and the results were subsequently analysed in terms of relative Marcus theory [22,23]. One difficulty encountered in this type of analysis was that few electron-transfer self-exchange rate constants had been reported for metalloproteins (cytochrome *c* was a notable exception [24]). To avoid this problem, the rate constant obtained for the reaction of the protein with the small reagent (the cross-reaction rate constant, k_{12}) was used with the self-exchange rate constant for the small reagent in question, k_{22}, to calculate the apparent self-exchange rate constant for the protein. These calculations were refined further to account for contributions of electrostatic interactions between the protein and the small molecule reagent to formation of the precursor complex and to dissociation of the successor complex [22,23]. The resulting electrostatics-corrected self-exchange rate constant, k_{11}^{corr}, reflects the reactivity of the protein in its reaction with the reagent in question. The term k_{11}^{corr} is only considered to be the true self-exchange rate constant of the protein if the same value, within an order of magnitude, is observed for the reaction of the protein with a variety of small reagents. Although Marcus theory predicts that protein self-exchange rate constants calculated from the reaction of a protein with a variety of inorganic reagents would be within an order of magnitude of each other, this is rarely found to be the case.

Experimental results obtained for nearly all proteins studied produced self-exchange rate constants for each protein that depended on the identity of the small molecule reagent. This inability of Marcus theory to predict the

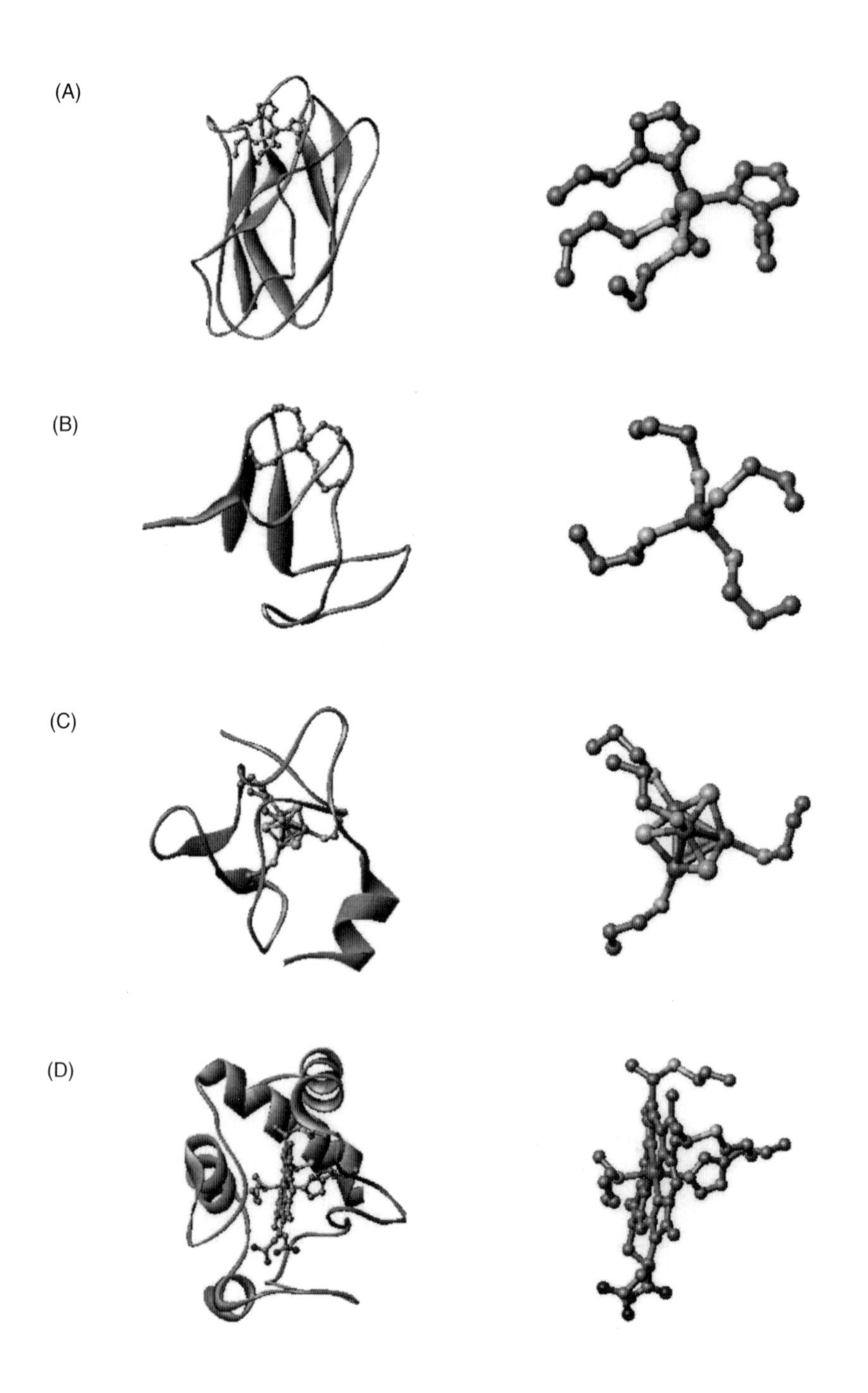

Figure 6. Structures of some simple electron-transfer proteins
The protein structures are shown on the left, and the corresponding co-ordination environment of the metal centre is enlarged on the right: (**A**) poplar plastocyanin, (**B**) rubredoxin *(Pyrococcus furiosus*), (**C**) high-potential iron protein (*Rhodocyclus tenus*), and (**D**) horse heart cytochrome *c*.

kinetic behaviour of metalloproteins in these reactions presumably arose because, in metalloproteins, access of the small molecules to the metal centres is somewhat restricted. Most notably, the precursor complex formed by the protein and the small reagent is likely to involve additional stabilizing interactions that are not present in reactions of two small molecules and that are, therefore, not accounted for by standard Marcus theory. In addition, if the ligands of the small molecule reagent are hydrophilic rather than hydrophobic in nature, then it is reasonable to expect that the precursor complex can be represented adequately by what might be considered a hard-sphere model. In such a model, the reagent and the protein are not intimately associated, and the distance separating the metal centre of the reagent and the metal centre of the protein is dominated by the distance from the metal atom at the active site of the protein to the protein surface.

These observations led to the conclusion that a protein in which the metal centre is relatively remote from the surface of the protein (kinetically inaccessible) will exhibit a relatively wide range in the values for the self-exchange rate constants calculated from the cross-reactions of the protein with a structurally diverse range of small molecule reagents. On the other hand, proteins that possess metal centres at or near the protein surface (kinetically accessible) will exhibit self-exchange rate constants that are independent of the nature of the small molecule reagent used in the cross-reaction. Proteins of the latter type, in fact, obey Marcus theory, but they are rare. The only protein reported to exhibit this type of behaviour is the blue copper protein stellacyanin [23,25]. Although the three-dimensional structure of stellacyanin had not been determined at the time that these electron-transfer studies were reported, the recently reported structure of this protein has, in fact, demonstrated that the blue copper site of stellacyanin is fully exposed to solvent [26]. For the majority of proteins that exhibit reactivity with small molecule reagents at variance with relative Marcus theory, the kinetic accessibility argument was eventually given a quantifiable basis that permitted the electron-transfer rate information obtained from such studies to provide an estimate of the distance over which electron transfer occurs [25].

Related studies by Tollin, Cusanovich and their colleagues studied the kinetics by which flavins reduce a variety of electron-transfer proteins [27,28] by employing flash photolysis [29] to exploit the chemistry first used with proteins by Massey and Palmer [30]:

$$\mathrm{Fl_{ox}} \xrightarrow{h\nu} {}^{3}\mathrm{Fl} \tag{13}$$

$${}^{3}\mathrm{Fl} + \mathrm{EDTA} \longrightarrow \mathrm{Fl}^{\bullet -} \tag{14}$$

$$\mathrm{Fl}^{\bullet -} + \mathrm{protein}^{\mathrm{ox}} \longrightarrow \mathrm{Fl_{ox}} + \mathrm{protein}^{\mathrm{red}} \tag{15}$$

In this scheme, Fl_{ox} represents the flavin hydroquinone, which is photoreduced to the triplet state. The flavin triplet reacts promptly with the sacrificial electron donor EDTA to form the semiquinone radical anion $Fl^{\bullet -}$, a strong reducing agent that reduces the electron-transfer protein. These studies demonstrated that the rate constants observed for electron transfer from various flavins to various metalloproteins are a function of the thermodynamic driving force (equilibrium constant) for the reaction, as is true for the reactions with inorganic complexes. In addition, these authors were able to correlate the relative reactivity of the flavins employed with the electrostatic properties of the flavins and with selected elements of their structure [27,28].

Refinements of metalloprotein electron-transfer experiments

A fundamental constraint of studies concerning a bimolecular reaction of an electron-transfer protein with either an inorganic complex or a flavin is that the relative positions of the electron donor and electron acceptor sites at the time of electron transfer are usually not known. In other words, the structure of the precursor complex is undefined. In principle, this limitation can be overcome by combining the electron donor and acceptor centres into a single molecular species and, thereby, converting the bimolecular reaction into a unimolecular reaction. The simplest means of implementing this strategy is to modify a simple electron-transfer protein for which the three-dimensional structure is known by introduction of a second oxidation–reduction centre somewhere on the surface of the protein. The ability to construct modified proteins of this type provides, in principle, the ability to vary the distance between the electron donor and acceptor centres and to vary the structure of the protein (the so-called 'intervening matter') situated between the electron donor and acceptor centres. These structural considerations comprise the primary determinants of the electronic coupling between the electron donor and acceptor centres, as described by Marcus theory. As a result, the appeal of producing electron-transfer proteins modified in this manner was substantial.

The principal challenge to achieving this goal was the development of specific protein-modification reagents and techniques. The first methods developed were based on the work of Matthews et al., who were interested in modification of surface histidyl residues of ribonuclease and other small proteins with ruthenium complexes [31]. The luminescent properties of the resulting ruthenium-modified histidyl residues vary with their environment so that they were useful as spectroscopic reporter groups of protein conformation. Using this approach, the groups of Isied [32] and Gray [33] modified horse heart cytochrome *c* with $[Ru(NH_3)_5H_2O]^{2+}$ to introduce ruthenium centres at surface histidyl residues of this protein. As part of this work, they also developed methods for purification of the reaction products and analysed the modified protein by HPLC tryptic-peptide-mapping techniques to establish the site of protein modification. Isied's group studied the electron-transfer properties of the modified cytochrome by the method of pulse radiolysis [34], whereas Gray's group used flash photolysis [29].

Pulse radiolysis requires access to a source of hydrated electrons, which are usually generated by a linear accelerator [34]. Through use of appropriate radical-quenching agents, it is possible to control the reactivity of the numerous free-radical species generated during such experiments so that the intramolecular electron-transfer reactions of the modified protein can be studied without interference of potential side reactions. Whereas this group and others (notably McLendon and Miller [35] and Salmon and Sykes [34]) have used pulse radiolysis in studies of metalloprotein electron-transfer kinetics, laser flash photolysis is more frequently used because it does not require a source of hydrated electrons and, therefore, is somewhat more widely available. Although the photochemical strategies used in these studies are reminiscent of the method of Massey and Palmer (eqns. 13–15), the flexibility provided by various photochemical approaches has provided a greater range of mechanistic insights, as reviewed by Scott [36]. Through application of these strategies and development of co-ordination complexes with ligands that modify the reduction potentials of the pendant electron-transfer centre, it has been possible to assess experimentally the magnitude of the contribution made by the thermodynamic driving force on the rate of electron transfer and to study the kinetics of both oxidation and reduction of the native electron-transfer centres of various proteins.

Interpretation of kinetics results obtained from these studies has been facilitated by the application of theoretical considerations outlined above. The relatively large distances separating the electron donor and acceptor centres in these modified proteins means that the probability that the transition state will proceed to products rather than revert to the reactant species is relatively low or, in other words, such reactions are non-adiabatic. For this reason, particular attention has been directed towards quantifying the contributions made by the two structural factors that are most important in determining the electronic coupling between the donor and acceptor centres: the distance separating the electron donor and acceptor and the structure of the protein located between the electron donor and acceptor centres. The most useful experimental information to use for this purpose is derived from kinetic analysis of a family of modified proteins in which a single modifying reagent has been used to produce a series of modified derivatives of a single protein in which various surface residues have been modified individually. In this way, the structural, reorganizational and electrochemical properties of the donor and acceptor centres can be held constant while the structural parameters of interest are varied. Whereas several proteins have now been studied in this manner, the greatest amount of information is available for modified derivatives of cytochrome *c* and myoglobin, primarily through the work of Gray and Winkler [37].

New challenges concerning intramolecular electron transfer

The availability of kinetics data for intramolecular electron-transfer reactions between electron donor and acceptor centres at defined locations in a protein environment raised new questions concerning the manner in which structural elements of proteins may or may not facilitate the electronic coupling of these centres and thereby influence the rate of electron transfer. How efficient is electronic coupling through a σ bond compared with coupling through a hydrogen bond or through space? Does the presence of aromatic groups between the donor and acceptor centres enhance coupling? Does the nature of the protein structure between the donor and acceptor centres matter, or is the distance separating the donor and acceptor sites the dominant factor? These questions reflect the single theoretical issue of whether or not the structure of the protein between the electron donor and acceptor centres (the 'intervening matter' or 'bridge') can be represented adequately as a medium of uniform dielectric properties [6] or if a more sophisticated description is required.

To address these issues, theoreticians have proposed modelling strategies to approximate the electronic coupling of electron donor and acceptor centres through protein structures. The most widely used of these strategies was proposed by Beratan, Betts and Onuchic [38]. The approach to defining the pathway of electron transfer used by these authors involved analysis of the structure of the protein that resides between the donor and acceptor sites and identification of possible structural elements through which electron transfer might occur (i.e. that could contribute to electronic coupling). The structural elements initially accounted for in this analysis included covalent bonds, hydrogen bonds and through-space (van der Waals) contacts. The most effective coupling was assumed to occur through covalent bonds, the least effective coupling through space. The relevant portion of the protein structure was then searched with the assistance of these criteria to identify the most effective 'pathway' for electron transfer. Refinements to this highly parameterized strategy have been described [39]. An alternative approach to electron-transfer pathway analysis developed by Siddarth and Marcus [40] attempted to estimate electronic coupling from first principles. In this strategy, an artificial intelligence algorithm was developed to identify the residues located between the donor and acceptor centres that were most likely to be involved in donor–acceptor coupling. A quantum-mechanical calculation of the electronic coupling matrix was then undertaken for this pathway with extended Hückel theory. In principle, an advantage of the Siddarth–Marcus approach is that it can account for the presence of π as well as σ bonds in the pathway.

These modelling approaches to interpreting intramolecular electron-transfer kinetics all assume that the structure of the modified proteins is adequately represented by the crystallographically determined atomic co-ordinates. Daizadeh and co-workers [41], however, have recently considered the consequences of protein structural dynamics on long-distance electron transfer through protein structures. These authors conclude that the effects of such structural fluctuations on electron-transfer kinetics increase as the distance between the donor and acceptor centres increases and that structural dynamics may influence the dependence of the electron-transfer rate constant on driving force.

As noted above, this interaction of theory and experiment has enhanced both perspectives of biological electron-transfer reactions. For example, one limitation of proteins such as cytochrome *c* and myoglobin for studies of this type is that the relatively irregular structures of these two proteins results in significantly different 'bridge' structures as the redox-active site introduced through chemical modification is moved to various positions on the surface of either protein. As a result, the ability of the experimentalist to vary the 'bridge' structure in a highly systematic fashion is restricted. Nevertheless, through judicious modification of selected copper and haem proteins, it has been possible to compare electronic coupling through helical and β-barrel sheet structure [42]. An alternative approach that was initiated by Dutton et al. [43] and McLendon et al. [44] involves the *de novo* design of synthetic electron-transfer proteins that possess predetermined structural motifs and possess the ability to incorporate oxidation–reduction centres.

Protein–protein electron-transfer reactions

Whereas intramolecular electron-transfer reactions are exhibited by a number of complex proteins that possess multiple oxidation–reduction centres, simpler proteins with just one such centre function by reacting with other electron-transfer proteins. Traditionally, such reactions have been studied under steady-state conditions. Whereas this approach can provide useful information, the uncertainties in rate constants determined in this fashion can be a limitation. To overcome this concern, stopped-flow spectroscopy has been used to study protein–protein electron-transfer reactions under conditions where the reaction is bimolecular. Technical challenges to such measurements are related to the overlap of the electronic absorption spectra of the reacting proteins and, frequently, the sensitivity of the oxidation state of one or both proteins to oxygen. However, these problems can usually be overcome through judicious selection of wavelengths to monitor reaction and through use of anaerobic techniques. Theoretical interpretation of bimolecular protein–protein electron-transfer kinetics requires consideration of the role of diffusion of the two reacting proteins through solution to produce the analogue of the precursor complex.

Work by Northrup and colleagues has been particularly successful in simulating bimolecular electron-transfer kinetics through the application of Brownian dynamics theory [45,46]. This approach involves the combined use of the Smoluchowski equation to describe the diffusion of proteins through solution and Debye–Hückel theory to approximate the manner in which electrostatic attraction between reacting proteins directs the diffusion of the two proteins towards each other. Implementation of this approach involves the calculation of several thousand trajectories of one protein towards the other to simulate the formation of encounter complexes that the two proteins form immediately prior to transfer of an electron from one to the other. The electrostatic energy of stabilization of each complex is calculated, and the charged residues on the surfaces of the two proteins that interact with each other are identified. By defining the criteria for formation of a 'successful' encounter complex, i.e. a complex that leads to electron transfer, it has been possible to account reasonably for the experimentally determined dependence of such reactions on both ionic strength and pH. For the reaction of cytochrome *c* with cytochrome *c* peroxidase, Brownian dynamics simulations led to the proposal of a primary and a secondary binding site for the cytochrome on the surface of the peroxidase [46]. Subsequently, the existence of low- and high-affinity binding sites was verified by fluorescence-quenching [47] and potentiometric [48] titrations. The functional significance of these binding sites in electron transfer from the cytochrome to the peroxidase is currently a topic of considerable experimental effort.

An alternative to studying protein–protein electron-transfer reactions under bimolecular conditions is to study such reactions under experimental conditions where the reaction is unimolecular. In such studies, a solution containing the two proteins of interest is prepared at low ionic strength and at a pH where complex formation is optimal. Electron transfer is then initiated by selective photoreduction of one of the proteins. Subsequent transfer of the electron to the other protein within the preformed complex is monitored spectrophotometrically. Photo-initiation of such reactions is generally achieved through the use of a flavin to selectively reduce one of the proteins or through use of a metal-substituted protein derivative. The latter approach has most frequently involved the use of a zinc-substituted haem protein as one member of the protein–protein complex because zinc-substituted haem proteins are readily converted to the highly reducing ^{3}Zn state by irradiation. This strategy has been employed most extensively by Hoffman and co-workers in characterization of the kinetics of electron transfer from cytochrome *c* to cytochrome *c* peroxidase [47]. These studies have detected a number of rate processes of similar but distinguishable magnitude (e.g. [47,49]) that have been interpreted in terms of multiple binding sites for the cytochrome on the surface of the peroxidase and multiple orientations of the two proteins with respect to each other within the preformed complex. Such conclusions are at least qualitatively consistent with the Brownian dynamics simulation of the reaction of these two proteins by Northrup and co-workers. Beratan and co-workers have subsequently applied their electron-transfer pathways algorithm to this system to prepare so-called electron-transfer contact maps of the peroxidase that predict those regions of the protein surface through which the most efficient electronic coupling to the haem iron centre may be achieved [49,50]. Remarkably, the surface regions of cytochrome *c* peroxidase predicted to exhibit such efficiency correspond reasonably well to the regions predicted to exhibit greatest contact frequency with cytochrome *c* from Brownian dynamics simulations.

An outlook

With improved theoretical methods to interpret and predict rate constants for protein-based electron-transfer reactions and improved strategies for genetic and chemical modification or design of proteins, new implications of biological electron-transfer reactions can be anticipated. Recent work in the author's laboratory, for example, has applied alternative strategies of introducing new oxidation–reduction centres to the surfaces of electron-transfer proteins that have led to improved insight concerning the properties of flavin-centred electron-transfer events [51]. Whereas considerably more remains to be learned about the behaviour of flavins in electron-transfer reactions, the electron-transfer properties of molybdopterins, dinuclear metal centres and other types of redox-active site have yet to be explored in depth. Electron-transfer reactions may also be used to probe other functional properties of proteins. Work by Gray and colleagues, for instance, has involved photoreduction of cytochrome *c* to initiate protein folding that provides new insights concerning the kinetics and mechanism of this process [52]. Photolytic activation of metalloenzymes and the electrochemical control of metalloenzyme oxidation state-linked activity are other eventual applications of this knowledge that will lead to new chemical insights and practical applications.

Summary

- *A wide range of biological processes makes extensive use of electron-transfer reactions.*
- *Rigorous characterization of a biological electron-transfer reaction requires a combination of kinetic, thermodynamic, structural and theoretical methods.*
- *The rate of electron transfer from an electron donor to an electron acceptor through a protein is dependent on the difference in reduction potential of the electron acceptor and electron donor and the distance over which electron transfer occurs. The manner in which the rate of electron transfer also depends on the structure of the protein located between the electron donor and acceptor sites remains an active topic of investigation.*

- *Diverse protein-engineering strategies are providing new insights into fundamental mechanistic considerations regarding electron-transfer properties of biological molecules, and they can provide novel means by which insights concerning biological electron-transfer reactions can be employed to develop new and useful types of chemistry.*

I thank Dr. Susanne Döpner for preparing Figure 6. Research in the author's laboratory is supported by Medical Research Council of Canada Operating Grants MT-7182 and MT-14021.

References

1. Marcus, R.A. (1996) Electron transfer reactions in chemistry. Theory and experiments. In *Protein Electron Transfer* (Bendall, D.S., ed.), pp. 249–272, BIOS Scientific Publishers, Oxford
2. Taube, H. (1970) *Electron transfer reactions of complex ions in solution*, Academic Press, New York
3. Taube, H. (1984) Electron transfer between metal complexes: retrospective. *Science* **226**, 1028–1036
4. Marcus, R.A. & Sutin, N. (1985) Electron transfers in chemistry and biology. *Biochim. Biophys. Acta* **811**, 265–322
5. Devault, D. (1984) *Quantum-Mechanical Tunnelling in Biological Systems*, Cambridge University Press, Cambridge
6. Moser, C.C., Keske, J.M., Warncke, K., Farid, R.S. & Dutton, P.L. (1992) Nature of biological electron transfer. *Nature (London)* **335**, 796–802
7. Moser, C.C. & Dutton, P.L. (1996) Outline of theory of protein electron transfer. In *Protein Electron Transfer* (Bendall, D.S., ed.), pp. 1–21, BIOS Scientific Publishers, Oxford
8. Newton, T.W. (1968) The kinetics of oxidation–reduction reactions. *J. Chem. Educ.* **45**, 571–575
9. Sutin, N. (1962) Electron exchange reactions. *Annu. Rev. Nuclear Sci.* **12**, 285–328
10. Heineman, W.R., Anderson, C.W., Halsall, H.B., Hurst, M.M., Johnson, J.M., Kreishman, G.P., Norris, B.J., Simone, M.J. & Su, C.-H. (1982) Studies of biological redox systems by thin-layer electrochemical techniques. In *Electrochemical and Spectrochemical Studies of Biological Redox Components* (Kadish, K.M., ed.), pp. 1–21, Advances in Chemistry Series, vol. 201, American Chemical Society, Washington DC
11. Taniguchi, V.T., Sailasuta-Scott, N., Anson, F.C. & Gray, H.B. (1980) Thermodynamics of metalloprotein electron transfer reactions. *Pure Appl. Chem.* **52**, 2275–2281
12. Yeh, P. & Kuwana, T. (1977) Reversible electrode reaction of cytochrome *c*. *Chem. Lett.* 1145–1148
13. Eddowes, M.J. & Hill, H.A.O. (1977) Novel method for the investigation of the electrochemistry of metalloproteins: cytochrome *c*. *J. Chem. Soc. Chem. Commun.* 771–772
14. Hagen, W.R. (1989) Direct electron transfer of redox proteins at the bare glassy carbon electrode. *Eur. J. Biochem.* **182**, 523–530
15. Barker, P.D. & Mauk, A.G. (1992) pH-Linked conformational regulation of a metalloprotein oxidation–reduction equilibrium: electrochemical analysis of the alkaline form of cytochrome *c*. *J. Am. Chem. Soc.* **114**, 3619–3624
16. Bond, A.M., Hill, H.A., Page, D.J., Psalti, I.S., Walton, N.J. (1990) Evidence for fast and discriminatory electron transfer of proteins at modified gold electrodes. *Eur. J. Biochem.* **191**, 737–742
17. Heinz, J. (1984) Cyclic voltammetry: 'Electrochemical spectroscopy'. *Angew. Chem. Int. Ed. Engl.* **23**, 831–847
18. Bond, A.M. (1994) Chemical and electrochemical approaches to the investigation of redox reactions of simple electron transfer metalloproteins. *Inorg. Chim. Acta* **226**, 293–340
19. Armstrong, F.A. (1991) Probing metalloproteins by voltammetry. *Struct. Bonding* **72**, 135–221

20. Armstrong, F.A. (1991) Voltammetry of metal centres in proteins. *Perspect. Bioinorg. Chem.* **1**, 141–182
21. Bennett, L.E. (1973) Metalloprotein redox reactions. *Prog. Inorg. Chem.* **18**, 1–176
22. Wherland, S. & Gray, H.B. (1976) Metalloprotein electron transfer reactions: analysis of reactivity of horse heart cytochrome *c* with inorganic complexes. *Proc. Natl. Acad. Sci. U.S.A.* **73**, 2950–2954
23. Wherland, S. & Gray, H.B. (1976) Electron transfer mechanisms employed by metalloproteins. In *Biological Aspects of Inorganic Chemistry* (Addison, A.W., Cullen, W., James, B.R. & Dolphin, D., eds.), pp. 289–368, Wiley-Interscience, New York
24. Gupta, R.K. & Redfield, A.G. (1970) Double nuclear magnetic resonance observation of electron exchange between ferri- and ferrocytochrome *c*. *Science* **169**, 1204–1206
25. Mauk, A.G., Scott, R.A. & Gray, H.B. (1980) Distances of electron transfer to and from metalloprotein redox sites in reactions with inorganic complexes. *J. Am. Chem. Soc.* **102**, 4360–4363
26. Hart, P.J., Nersissian, A.M., Hermann, R.G., Nalbandyan, R.M. & Valentine, J.A. (1996) A missing link in cupredoxins: crystal structure of cucumber stellacyanin at 1.6 Å resolution. *Protein Sci.* **5**, 2175–2183
27. Cusanovich, M.A. (1991) Photochemical initiation of electron transfer reactions. *Photochem. Photobiol.* **53**, 845–857
28. Cusanovich, M.A. & Tollin, G. (1996) Kinetics of electron transfer of *c*-type cytochromes with small reagents. In *Cytochrome c. A Multidisciplinary Approach* (Scott, R.A. & Mauk, A.G., eds.), pp. 489–513, University Science Books, Sausilito
29. Sawicki, C.A. & Morris, R.J. (1981) Flash photolysis of hemoglobin. *Methods Enzymol.* **76**, 667–681
30. Massey, V. & Palmer, G. (1966) On the existence of spectrally distinct classes of flavoprotein semiquinones. A new method for the quantitative production of flavoprotein semiquinones. *Biochemistry* **5**, 3181–3189
31. Matthews, C.R., Erickson, P.M. & Froebe, C.L. (1980) The pentaammineruthenium(III)-histidine complex in ribonuclease A as an optical probe of conformational change. *Biochim. Biophys. Acta* **624**, 499–510
32. Isied, S.S., Worosila, G. & Atherton, S.J. (1982) Electron transfer across polypeptides. 4. Intramolecular electron transfer from ruthenium(II) to iron(III) in histidine-33 modified horse heart cytochrome *c*. *J. Am. Chem. Soc.* **104**, 7659–7661
33. Yocom, K.M., Shelton, J.B., Shelton, J.R., Schroeder, W.A., Worosila, G., Isied, S.S., Bordignon, E. & Gray, H.B. (1982) Preparation and characterization of a pentaammineruthenium(III) derivative of horse heart ferricytochrome *c*. *Proc. Natl. Acad. Sci. U.S.A.* **79**, 7052–7055
34. Salmon, G.A. & Sykes, A.G. (1993) Pulse radiolysis. *Methods Enzymol.* **227**, 522–534
35. McLendon, G. & Miller, J.R. (1985) The dependence of biological electron transfer rates on exothermicity: the cytochrome *c*/cytochrome b_5 couple. *J. Am. Chem. Soc.* **107**, 7811–7816
36. Scott, R.A. (1996) Long-range intramolecular electron transfer reactions in cytochrome *c*. In *Cytochrome c. A Multidisciplinary Approach* (Scott, R.A. & Mauk, A.G., eds.), pp. 515–541, University Science Books, Sausilito
37. Gray, H.B. & Winkler, J.R. (1996) Electron transfer in proteins. *Annu. Rev. Biochem.* **65**, 537–561
38. Beratan, D.N., Betts, J.N. & Onuchic, J.N. (1991) Protein electron transfer rates set by the bridging secondary and tertiary structure. *Science* **252**, 1285–1288
39. Beratan, D.N. & Onuchic, J.N. (1996) The protein bridge between redox centres. In *Protein Electron Transfer* (Bendall, D.S., ed.), pp. 23–42, BIOS Scientific Publishers, Oxford
40. Siddarth, P. & Marcus, R.A. (1993) Electron-transfer reactions in proteins: an artificial intelligence approach to electronic coupling. *J. Phys. Chem.* **97**, 2400–2405
41. Daizadeh, I., Medvedev, E.S. & Stuchebrukhov, A.A. (1997) Effects of protein dynamics on biological electron transfer. *Proc. Natl. Acad. Sci. U.S.A.* **94**, 3703–3708
42. Gray, H.B. & Winkler, J.R. (1997) Electron tunneling in structurally engineered proteins. *J. Electroanal. Chem.* **438**, 43–47

43. Robertson, D.E., Farid, R.S., Moser, C.C., Urbauer, J.L., Mulholland, S.E., Pidikiti, R., Lear, J.D., Wand, A.J., DeGrado, W.F. & Dutton, P.L. (1994) Design and synthesis of multi-haem proteins. *Nature (London)* **368**, 425–432
44. Mutz, M.W., McLendon, G.L., Wishart, J.F., Gaillard, E.R. & Corin, A.F. (1996) Conformational dependence of electron transfer across *de novo* designed metalloproteins. *Proc. Natl. Acad. Sci. U.S.A.* **93**, 9521–9526
45. Northrup, S.H. (1996) Computer modelling of protein-protein interactions. In *Protein Electron Transfer* (Bendall, D.S., ed.), pp. 69–97, BIOS Scientific Publishers, Oxford
46. Northrup, S.H., Boles, J.O. & Reynolds, J.C.L. (1988) Brownian dynamics of cytochrome *c* and cytochrome *c* peroxidase association. *Science* **241**, 67–70
47. Stemp, E.D.A. & Hoffman, B.M. (1993) Cytochrome *c* peroxidase binds two molecules of cytochrome *c*: evidence for a low-affinity, electron-transfer-active site on cytochrome *c* peroxidase. *Biochemistry* **32**, 10848–10865
48. Mauk, M.R., Ferrer, J.C. & Mauk, A.G. (1994) Proton linkage in formation of the cytochrome *c*-cytochrome *c* peroxidase complex: electrostatic properties of the high- and low-affinity cytochrome binding sites on the peroxidase. *Biochemistry* **33**, 12609–12614
49. Nocek, J.M., Zhou, J.S., De Forest, S., Priyadarshy, S., Beratan, D.N., Onuchic, J.N. & Hoffman, B.M. (1996) Theory and practice of electron transfer within protein-protein complexes. Application to the multidomain binding of cytochrome *c* by cytochrome *c* peroxidase. *Chem. Rev.* **96**, 2459–2489
50. Skourtis, S.S. & Beratan, D.N. (1997) Electron transfer contact maps. *J. Phys. Chem.* **B101**, 1215–1234
51. Twitchett, M.B., Ferrer, J.C., Siddarth, P. and Mauk, A.G. (1997) Intramolecular electron transfer kinetics of a synthetic flavocytochrome *c*. *J. Am. Chem. Soc.* **119**, 435–436
52. Pascher, T., Chesick, J.P., Winkler, J.R. & Gray, H.B. (1996) Protein folding triggered by electron transfer. *Science* **271**, 1558–1560

8

Molybdenum enzymes

Russ Hille

Department of Medical Biochemistry, The Ohio State University, Columbus, OH 43210-1218, U.S.A.

Introduction

Molybdenum-containing enzymes are found throughout the biological world, and catalyse critical reactions in the metabolism of purines and aldehydes, as well as nitrogen- and sulphur-containing compounds. The reason molybdenum is found in the active sites of so many enzymes has to do with the fact that molybdate salts are extremely water soluble; although the ratio of iron to molybdenum in the earth's crust is 3000:1 in favour of iron, molybdenum is approximately 5-fold more abundant in sea water owing to the greater solubility of molybdates relative to iron oxides. Coupled to this property of solubility are several aspects of molybdenum chemistry that make the metal well suited to participate in the catalysis of certain types of biologically important reaction. The metal is found in three different oxidation states (IV, V and VI) and has the desirable ability to couple biological compounds that are obligatory two-electron carriers (e.g. NADH) with obligatory one-electron carriers (e.g. iron–sulphur centres and cytochromes). In addition, reduction of molybdenum(VI) is tightly coupled to protonation of ligands to the metal, typically Mo=O or Mo=S groups to give Mo–OH or Mo–SH, respectively. Finally, Mo=O groups are found to be surprisingly labile, and a number of chemical systems are known in which the metal centre is able to act as a catalyst for oxygen atom transfer to and from suitable acceptors and donors. Given the ready availability of molybdenum in the environment and the unique aspects of its chemistry, it is hardly surprising that enzymes have evolved which utilize molybdenum to catalyse a variety of metabolic reactions.

Classification of molybdenum enzymes

There are at present more than 50 known molybdenum-containing enzymes (not counting enzymes that catalyse the same reaction in different species) [1]. In nitrogenase, molybdenum is found in a unique multinuclear metal cluster of complex structure. All other molybdenum-containing enzymes have active sites possessing a single molybdenum ion, with the metal co-ordinated to at least one equivalent of an unusual pterin-containing cofactor that has the structure shown in Figure 1. This cofactor, called pyranopterin here (abbreviated to ppt; other names found in the literature include 'molybdopterin', 'molybdenum cofactor' or simply 'moco'), consists of a pterin species fused to a pyran ring that has a dithiolene unit by which it co-ordinates to the metal. This same cofactor is also found in all tungsten-containing enzymes that have been examined to date [2]. In molybdenum enzymes from eukaryotic sources the cofactor consists simply of the structure shown in Figure 1, but in bacterial and archaeal enzymes it is found as a dinucleotide condensed with cytosine, guanine, adenine or hypoxanthine nucleotides. Not surprisingly, the biosynthesis of this cofactor is rather complex, involving at least 10 discrete gene products. These can be grouped loosely into proteins responsible for the transport of molybdate, the synthesis of the heterocycle, the incorporation of the dithiolene sulphur and, finally, the elaboration to the dinucleotide (in the case of the bacterial enzymes) and incorporation into the apoenzyme. Genes involved in one or more of these processes have been identified in *Escherichia coli*, yeast, *Drosophila*, *Arabidopsis* and humans, with considerable amino acid sequence similarity to be found among cognate proteins from different species [3].

The molybdenum enzymes requiring the ppt cofactor fall into three distinct families on the basis of the structures of their molybdenum centres [1]. The first of these consists of enzymes that have an active site consisting of a

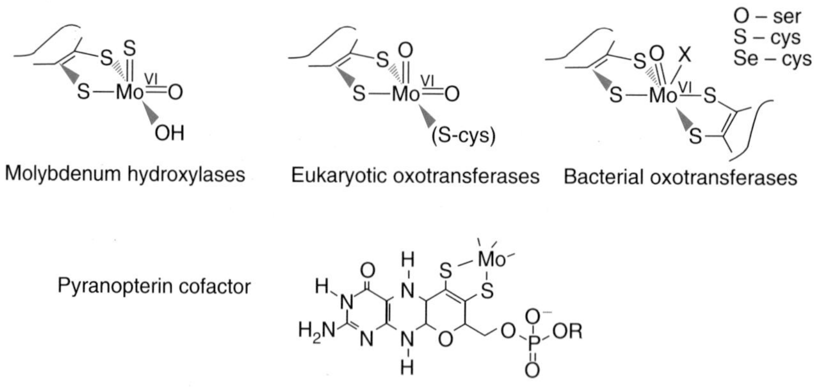

Molybdenum hydroxylases | Eukaryotic oxotransferases | Bacterial oxotransferases

Pyranopterin cofactor

Figure 1. The three families of molybdenum enzymes and the structure of the pterin cofactor

The structures drawn are for the oxidized Mo(VI) forms of each enzyme.

(ppt)MoOS(OH) unit with a single equivalent of ppt co-ordinated to molybdenum in the oxidized enzyme (Figure 1). This group of enzymes includes xanthine oxidase and xanthine dehydrogenase, as well as a number of aldehyde oxidoreductases (including three distinct aldehyde oxidases in humans), and a variety of enzymes that catalyse the hydroxylation of pteridines, quinazolines and related compounds. All these enzymes catalyse the hydroxylation of an activated carbon centre, in either an aromatic heterocycle or the carbonyl of an aldehyde, in a reaction that formally involves the removal of hydride from substrate and its replacement with hydroxide. The second family of enzymes consists of the sulphite oxidases and nitrate reductases from a variety of eukaryotic organisms; these enzymes have a (ppt)MoO_2(S–Cys) unit in their active sites (Figure 1) and they catalyse oxygen atom transfer to or from their substrates in reactions that are mimicked very well in the inorganic systems discussed further below. Finally, a diverse group of bacterial and archaeal enzymes exists, containing a molybdenum centre with the overall stoichiometry $(ppt)_2$MoOX in their active sites. Two equivalents of the ppt cofactor are co-ordinated to the molybdenum, along with an oxo group (or hydroxide, in some cases) and an additional ligand (X), which is either a serine, a cysteine or even a selenocysteine residue. Most of these enzymes catalyse oxygen atom-transfer reactions, although some appear to catalyse simple dehydrogenation or oxidation–reduction reactions. I will now consider each of these three families in turn, with specific reference to their mechanisms of action.

The molybdenum hydroxylases

The molybdenum hydroxylases are complex enzymes, invariably possessing several redox-active centres in addition to the molybdenum. The hydroxlyation reaction occurs at the molybdenum site, whereas the other redox-active centres serve to transfer electrons to an external electron acceptor, such as O_2 or NAD. The focus of this Chapter is the molybdenum centre. Electron transfer involving the other redox cofactors will not be discussed. The one member of this family for which a crystal structure exists is the aldehyde oxidoreductase from *Desulphovibrio gigas*, an α_2 dimeric enzyme with each subunit comprising two [2Fe–2S] centres of the spinach ferredoxin variety in addition to its molybdenum centre [4,5]. The crystal structure reveals that the two identical subunits of the enzyme are built up in a modular fashion: two discrete domains at the N-terminus of the enzyme are responsible for binding the two iron–sulphur centres, with two larger domains together constituting the molybdenum-binding portion of the protein. The molybdenum centre is buried deeply in the enzyme, with a tunnel ~15 Å long providing solvent access to the metal site. The molybdenum centre possesses the overall structure shown in Figure 1, with the molybdenum co-ordination sphere consisting of terminal Mo=O and Mo=S groups and an Mo–OH or $MoOH_2$, in addition to the single equivalent of the ppt; the overall co-ordination geometry is that of a (distorted) square pyramid with the Mo=S at the apex. At present there is some ambiguity as to whether the second oxygen ligand is singly or doubly protonated, an issue that is important from a mechanistic standpoint as it is this oxygen that is incorporated into the hydroxyl group of product (see later). This ligand position points directly into the solvent access channel to the active site. Apart from the solvent access channel, the immediate protein environment about the metal is sterically constrained, particularly around the Mo=O. On the opposite side of the metal from the apical Mo=S group, a glutamate residue lies near (but not co-ordinated to) the metal. This residue may act as a general base in the course of the hydroxylation reaction, taking up a proton from the Mo–OH group during the reaction [5]. It is interesting that the amino group of the ppt is hydrogen-bonded to one of the cysteine ligands of the second iron–sulphur centre. Evidently, an important role of the cofactor in this enzyme is to mediate electron transfer from the molybdenum to the iron–sulphur centres. Also, given that there is no protein ligand to the metal, the cofactor serves to anchor the molybdenum centre in the polypeptide through numerous hydrogen-bonding interactions with amino acid residues. Because the pterin ring lies opposite the molybdenum from the solvent access channel, the pterin is not likely to participate directly in the chemistry catalysed by the enzyme.

Other molybdenum hydroxylases possess a flavin centre in addition to the three redox-active centres found in the *D. gigas* aldehyde oxidoreductase. From a comparison of the amino acid sequences for these proteins it is clear that a discrete flavin-binding domain has been inserted between the iron–sulphur and molybdenum-binding domains. This is a frequently encountered arrangement for molybdenum hydroxylases, although in some bacterial enzymes of this family the redox-active centres are found in separate subunits rather than as domains within a single polypeptide. In humans, in addition to xanthine oxidoreductase (which catalyses the hydroxylation of hypoxanthine to xanthine and xanthine to uric acid) there are at least three distinct aldehyde oxidases. There is a liver enzyme that is involved in generic aldehyde metabolism, another enzyme responsible for metabolizing retinal to retinoic acid in the retina, and a third enzyme that appears to catalyse the synthesis of retinoic acid as a developmental signal in the vertebrate embryo.

The reaction mechanism of the molybdenum hydroxylases has attracted a great deal of attention, principally due to the unique nature of the reaction catalysed. Other hydroxylating systems possess flavin, pterin, copper, haem or non-haem iron in their active sites, and catalyse hydroxylations according to the following stoichiometry (see Chapters 4–6 in this volume):

$$S + O_2 + 2[e^-] + 2H^+ \longrightarrow SOH + H_2O \quad (1)$$

However, the molybdenum hydroxylases catalyse the reaction in the following way:

$$S + H_2O \longrightarrow SOH + 2H^+ + 2[e^-] \quad (2)$$

Water rather than molecular oxygen is the source of the oxygen atom incorporated into product, and reducing equivalents are generated rather than consumed. The molybdenum hydroxylases thus avoid the formal spin-forbidden reaction of triplet oxygen with a singlet substrate, but are not able to take advantage of the compelling thermodynamic driving force for the hydroxylation reaction that is provided by forming hypervalent metal-oxo oxidation states during the reduction of dioxygen to water (see Chapters 4–6 in this volume).

Certainly, the molybdenum hydroxylase that is best understood from a mechanistic viewpoint is xanthine oxidase from cow's milk. It has been known for some time that this enzyme bears a labile oxygen that is the immediate source of the oxygen incorporated into product, which is regenerated at the completion of each catalytic sequence with oxygen derived from solvent. On the basis of studies in several laboratories, it appears that this catalytically labile oxygen is the Mo–OH group of the molybdenum centre [6–8]. Although

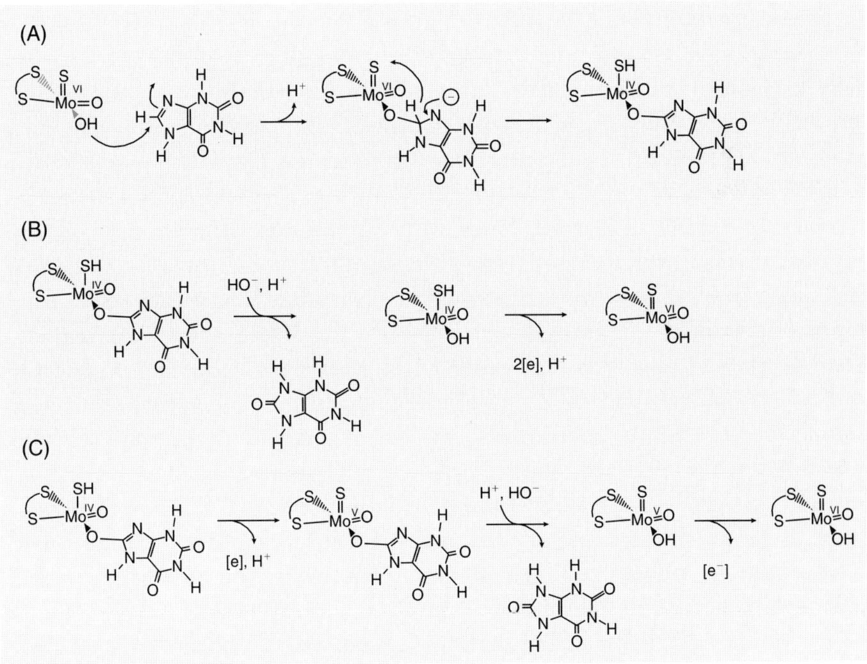

(A)

(B)

HO⁻, H⁺

2[e], H⁺

(C)

[e], H⁺

H⁺, HO⁻

[e⁻]

Figure 2. A proposed reaction mechanism of xanthine oxidase
(**A**) Reaction of the oxidized enzyme with substrate to give a reduced enzyme–product complex. (**B**) Decay of the reduced enzyme–product complex via product dissociation. (**C**) Decay of the reduced enzyme–product complex via re-oxidation prior to product release.

other mechanisms have been proposed (reviewed in [1]), the most reasonable chemistry leading to hydroxylation of substrate involves an initial nucleophilic attack by this Mo–OH group on the C-8 position of xanthine, followed by hydride transfer from C-8 to the Mo=S group (Figure 2), yielding Mo(IV) in complex with product. The attacking oxygen co-ordinated to the molybdenum is shown here as hydroxide rather than water, although at this point the issue is not settled. One would expect hydroxide to be a more effective nucleophile than water, and in the high Mo(VI) oxidation state the metal can act as a Lewis acid and favour unprotonated over protonated forms of oxygen (i.e. Mo=O and Mo–OH versus Mo–OH_2). Nevertheless, the Mo–O distance observed in the crystal structure of the *D. gigas* aldehyde oxidoreductase is 2.2 Å, more consistent with water than hydroxide.

Xanthine oxidase is able to hydroxylate a wide range of purines, pteridines and other aromatic heterocycles. The decay of the resulting reduced enzyme–product complex proceeds via two alternate pathways [1]. The relative contributions of these two pathways vary from one substrate to another. For xanthine and most other substrates the principal decay route involves initial dissociation of product from the molybdenum and its replacement with hydroxide from solvent (Figure 2). The resultant (ppt)Mo(IV)O(SH)(OH) complex then decays to the starting (ppt)Mo(VI)OS(OH) species as electrons are transferred to the other redox-active centres of the enzyme, with concomitant deprotonation of the Mo–SH group. For certain substrates the predominant pathway for decay of the reduced enzyme–product complex involves initial one-electron oxidation (again by electron transfer to the other redox-active centres in the enzyme) and deprotonation of the Mo–SH group to give a (ppt)Mo(V)OS(OR) species that can be detected by EPR spectroscopy [Mo(IV) and Mo(VI) are EPR-silent]. The EPR signal exhibited by this species has varying degrees of intensity with different substrates for the enzyme, and has been studied extensively [9]. No coupling of protons to the molybdenum EPR signal is observed (this is the basis for concluding that the Mo–SH group deprotonates in the course of forming the EPR-active species). However, when 8-[^{13}C]xanthine is used to generate the EPR signal, coupling is observed to the $I=1/2$ nucleus of the ^{13}C. The purine nucleus must therefore remain co-ordinated to the Mo(V) centre that gives rise to the EPR signal. Decay of this species is likely to proceed by dissociation of product, followed by co-ordination of hydroxide and, finally, full re-oxidation to Mo(VI). The pH dependence of each of these steps has been investigated (reviewed in [1]) and found to be consistent with the addition or loss of protons in the course of the overall hydroxylation reaction indicated in Figure 2(C). In the case of the first step of the reaction, evidence has been found for an essential active-site base (possibly Glu-1261 or the homologous Glu-689 identified above in the *D. gigas* aldehyde oxidoreductase structure), whose likely role is to deprotonate the Mo–OH, thereby making it a better nucleophile.

The eukaryotic molybdenum oxotransferases

The eukaryotic molybdenum oxotransferases also possess other redox-active centres in addition to their molybdenum centres. Sulphite oxidases from vertebrates have an N-terminal cytochrome *b* domain, followed by the molybdenum-binding portion of the enzyme. The enzyme oxidizes sulphite to sulphate at the molybdenum and subsequently reduces cytochrome *c*, a reaction that takes place at the *b*-type cytochrome. Nitrate reductase has a molybdenum-binding portion of the protein with considerable amino acid sequence similarity to that of sulphite oxidase. This is linked to a cytochrome *b* domain (highly homologous not only to that of sulphite oxidase, but also to bovine cytochrome b_5), and a flavin-binding domain. In this enzyme, reducing equivalents in the form of NADH or NADPH enter via the flavin, and are subsequently transferred to the molybdenum centre (presumably via the *b*-type cytochrome), where nitrate is reduced to nitrite. The molybdenum centres of these enzymes have long been recognized to be fundamentally very similar to each other, but distinct from those of xanthine oxidase and other molybdenum hydroxylases. The fact that one enzyme oxidizes sulphite and the other reduces nitrate has more to do with the relative stabilities of the N=O and S=O bonds than with intrinsic differences in the two active sites; this important point has been quantified thermodynamically [10].

The crystal structure of chicken sulphite oxidase has been reported recently [11], and the protein is found to consist of three distinct domains: the *b*-type cytochrome domain referred to above, a molybdenum-binding domain and a third domain possessing an overall fold of the immunoglobulin type that provides the subunit interface in the dimeric enzyme. The molybdenum centre has a distorted square-pyramidal co-ordination geometry, with an apical Mo=O; the ppt cofactor, an Mo–OH, and a co-ordinated cysteine define the base of the pyramid (Figure 1). Given the X-ray absorption spectroscopic evidence that the oxidized form of sulphite oxidase has two Mo=O groups, but that one becomes protonated to Mo–OH upon reduction of the enzyme [12], it is likely that the enzyme used for the crystallographic determination was reduced in the crystal, either by the X-ray beam or by contaminating sulphite in the $LiSO_4$ that was used to crystallize the protein. Adjacent to the molybdenum centre is a likely substrate-binding pocket that has three arginine residues in an appropriate position to bind the negatively charged sulphite ion.

The reaction catalysed by sulphite oxidase is fundamentally more straightforward than that of the molybdenum hydroxylases and, as alluded to above, there is abundant chemical precedent for reactions of the type shown in Figure 3. These dioxo Mo(VI) compounds can transfer an oxo group to a suitable acceptor (usually a phosphine), and the cognate mono-oxo Mo(IV) complexes accept an oxo group from a suitable oxo donor (e.g. DMSO) to return to the dioxo Mo(VI) form [13]. In the case of sulphite oxidase, the most likely mechanism involves attack of a sulphite lone pair on one of the oxo groups of the

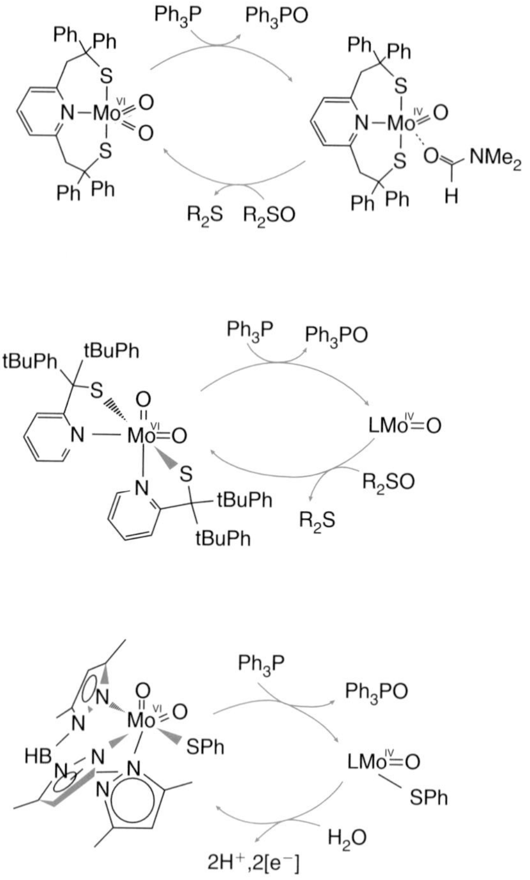

Figure 3. The model compounds related to the eukaryotic molybdenum oxotransferases and their reactions with oxygen atom acceptors and donors
Note that the third example is able to use water as the source of oxygen in regenerating the labile oxygen from the Mo(IV) complex, as does sulphite oxidase. Ph, phenyl; t-BuPh, t-butylphenyl; SPh, thiophenol; L, molybdenum ligand not including oxygen and SPh.

molybdenum centre (probably the equatorial rather than apical oxo group, as indicated in Figure 4). This gives a reduced Mo(IV)–product complex that must then decay by product dissociation and replacement by hydroxide to give the form of the enzyme that was apparently analysed crystallographically. As with the molybdenum hydroxylases, completion of the catalytic sequence involves re-oxidation via electron transfer (to the haem), with concomitant deprotonation of the Mo–OH group to regenerate the second oxo group.

Consistent with the first step in the above mechanism, in which the chemistry is initiated by a substrate lone pair rather than by co-ordination of sulphite to the molybdenum centre via one of its oxyanion groups, it has been found that dimethylsulphite is able to reduce sulphite oxidase [14]. The K_d for this reaction is ~500-fold greater than is seen with sulphite as substrate but,

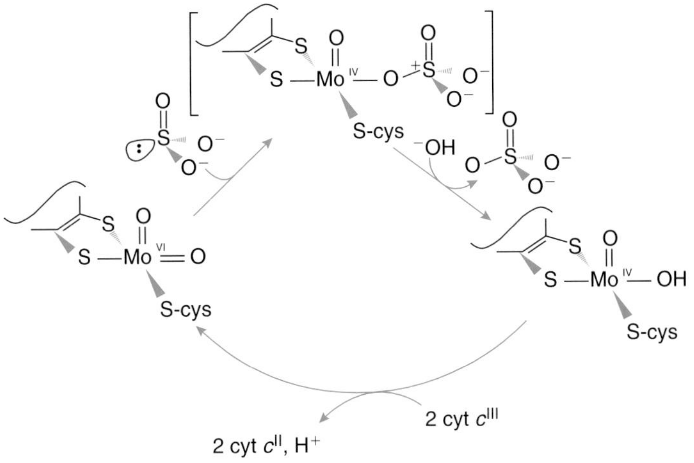

Figure 4. A plausible reaction mechanism for the reaction of sulphite oxidase

importantly, the limiting rate constant for enzyme reduction in the high substrate-concentration limit is essentially unchanged (170 s^{-1}, compared with 194 s^{-1} for sulphite). The implication is that blocking the substrate oxyanion groups by methylation has no effect on breakdown of the enzyme–substrate Michaelis complex.

The bacterial molybdenum oxotransferases and related enzymes

As indicated above, the bacterial oxotransferases constitute the most diverse family of the enzymes considered here. Structurally, these enzymes may have either serine, cysteine or selenocysteine co-ordinated to the metal, and it appears that a few of these enzymes possess Mo=S rather than Mo=O in their active sites. Mechanistically, some of these enzymes catalyse dehydrogenation or oxidation–reduction reactions rather than oxygen atom-transfer reactions (e.g. the several formate dehydrogenases from *E. coli* and polysulphide reductase from *Wolinella succinogenes*, respectively). In contrast to the molybdenum hydroxylases and eukaryotic oxotransferases, many of these enzymes do not contain any flavins or iron centres that would give strong absorbance, so that the molybdenum centre, which has distinct, but weak, spectral properties, is not masked. These enzymes therefore provide a unique opportunity for mechanistic studies. The structures of two enzymes from this family have been determined crystallographically: the DMSO reductase from *Rhodobacter* species (both *R. sphaeroides* and *R. capsulatus*) [15–17], and the formate dehydrogenase H from *E. coli* [18]. DMSO reductase has a serine co-ordinated to the molybdenum centre and no other redox-active centres are present, whereas formate dehydrogenase H has selenocysteine co-ordinated to the molybdenum and possesses a [4Fe–4S] centre as an additional redox group;

both enzymes utilize the guanosine dinucleotide form of the ppt cofactor. The overall protein folds are superficially similar, but do differ in important ways. In both cases the molybdenum co-ordination sphere of the active site consists of a pair of ppt cofactors, an oxygen ligand and the amino acid residue. The overall co-ordination geometry is that of a trigonal prism, as indicated in Figure 1.

There is actually considerable uncertainty regarding the active site structure of DMSO reductase. The crystal structure of the DMSO reductase from *R. capsulatus* has been determined independently by two different groups [16,17], and whereas the overall polypeptide fold is virtually identical to that of the *R. sphaeroides* enzyme, two alternate structures for the molybdenum centre, each different from that shown in Figure 1, have been reported. In the first [16], although there are two equivalents of the ppt cofactor present in the protein, only one is co-ordinated to the metal. The second cofactor, although close, is not co-ordinated to the metal; the two ligand co-ordination positions at which the second cofactor would otherwise bind are occupied by terminal Mo=O groups. The second structure shows both equivalents of the ppt co-ordinated to the metal, but a seventh ligand to the metal is observed that is modelled as a long Mo=O group [17]. It has not been clarified whether the enzyme from either of these crystals is completely native and enzymically active. It is hoped that this rather confusing state of affairs will soon be resolved as the crystal structures are further refined.

Making the assumption that the active site of oxidized DMSO reductase has the structure shown in Figure 1, a mechanism can be hypothesized in which the enzyme functions in a Ping Pong fashion between mono-oxo Mo(VI) and *des* oxo Mo(IV) forms, with the Mo=O oxygen of oxidized enzyme derived from substrate upon reaction with the reduced enzyme (consistent with the experimental observation that the enzyme possesses a catalytically labile oxygen in its molybdenum centre). It must be recognized, however, that given the uncertainties in the structure of the active site it is difficult to make any reliable predictions concerning the reaction mechanism at present.

Perspectives

Our understanding of the molybdenum enzymes is advancing rapidly, making it possible to address increasingly incisive questions concerning their structure and function. With regard to the reaction mechanism of the molybdenum hydroxylases, one key question that remains unresolved has to do with the structure of the initial Mo(IV)–P complex formed in the reaction with substrate: specifically, whether there is significant Mo–C bond character in this intermediate. It is hoped that the crystal structure of xanthine oxidase (or another member of this family that possesses a flavin domain in addition to the molybdenum and iron–sulphur centres common to all enzymes of this group) will be forthcoming soon. The development of suitable expression systems for these enzymes has become an extremely important objective, as these are essential for site-directed mutagenesis studies that can address the roles of specific amino acid residues in various steps of the catalytic sequence.

The recently reported crystal structure for sulphite oxidase provides a context for site-directed mutagenesis studies, and an expression system (for the human protein, at least) has been reported [19]. It is to be expected that rapid progress will be made in ascertaining the roles of specific amino acid residues in the reaction mechanism of this enzyme. Our understanding of the (plant) nitrate reductase is not nearly as well developed, although an expression system for the *Arabidopsis* enzyme has been reported recently [20]. Certainly the most important outstanding issue is the structure of the protein; in particular, the disposition of the molybdenum-, haem- and flavin-containing domains with respect to one another. Little mechanistic work has been done on any of the plant enzymes, and this represents an area that is likely to prove very attractive to new investigators.

For DMSO reductase, certainly the most important goal in the near term is resolution of the structure of the active site of the enzyme, but in the longer term, the extremely poor understanding of the reaction mechanism in general must be addressed. Because the only redox-active group in this enzyme is a molybdenum centre, there is an excellent opportunity to apply a wide range of spectroscopic techniques to probe not only the structures of oxidized and reduced enzyme, but also catalytic intermediates that are likely to be identified kinetically. The relationship of this enzyme to others with two equivalents of the ppt cofactor is presently only very poorly understood and, in particular, the differences in properties of a molybdenum centre with serine versus cysteine or selenocysteine co-ordination to the metal must be addressed. In addition, it remains to be seen whether all these enzymes are sufficiently closely related to justify their being classified in a single family (albeit with several subgroups). For all three families of molybdenum enzyme, it is expected that the application of computational approaches (including density functional calculations) will yield new insights into their reaction mechanisms.

Finally, our understanding of the biosynthesis of the ppt cofactor of the molybdenum (and tungsten) enzymes has advanced to the stage that most of the intermediates in the metabolic pathway and many of the gene products involved in the interconversion of these intermediates have been identified. Goals of immediate importance involve the expression, purification and characterization of these gene products so that the metabolic pathways in organisms ranging from *E. coli* to *Drosophila*, *Arabidopsis* and humans can be compared and contrasted. This area is advancing very rapidly at present, and it is to be expected that we will see important breakthroughs over the next few years.

Summary

- *There are many molybdenum-containing enzymes distributed throughout the biosphere. The availability of molybdenum to biological systems is due to the high water solubility of oxidized forms of the metal.*
- *Molybdenum enzymes can be grouped on the basis of the structure of the metal centre. Three principal families of enzyme exist, with active sites consisting of (ppt)MoOS(OH) (the molybdenum hydroxylases), (ppt)MoO_2(S–Cys) (the eukaryotic oxotransferases) and $(ppt)_2MoOX$ (the bacterial oxotransferases). Here, ppt represents a unique ppt cofactor (pyranopterin) that co-ordinates to the metal, and X is a metal-liganded serine, cysteine or selenocysteine.*
- *The molybdenum hydroxylases catalyse their reactions differently to other hydroxylase enzymes, with water rather than molecular oxygen as the ultimate source of the oxygen atom incorporated into product, and with the generation rather than consumption of reducing equivalents. The active sites possess a catalytically labile Mo–OH (or possibly Mo–OH_2) group that is transferred to substrate in the course of the hydroxylation reaction. These enzymes invariably have other redox-active centres.*
- *The eukaryotic oxotransferases consist of the sulphite oxidases and plant nitrate reductases. They catalyse the transfer of an oxygen atom to or from their substrate (to and from nitrate) in a manner that involves formal oxidation-state changes of the molybdenum. As with the molybdenum hydroxylases, the ultimate source of oxygen is water rather than molecular oxygen.*
- *The bacterial oxotransferases and related enzymes differ from the other two groups of molybdenum enzymes in having two equivalents of the ppt cofactor co-ordinated to the metal. This family is quite diverse, as reflected in the fact that serine, cysteine or selenocysteine may be found co-ordinated to the molybdenum, depending on the enzyme.*
- *As in the case of the molybdenum hydroxylases, both eukaryotic and bacterial oxotransferases utilize water (rather than molecular oxygen) as the source of the oxygen atom incorporated into product, although for these enzymes, the catalytically labile oxygen in the active site is an Mo=O group rather than an Mo–OH.*

References

1. Hille, R. (1996) The mononuclear molybdenum enzymes. *Chem. Rev.* **96**, 2757–2816
2. Rajagopalan, K.V. & Johnson, J.L. (1992) The pterin molybdenum cofactors. *J. Biol. Chem.* **267**, 10199–10202

3. Mendel, R.R. & Schwarz, G. (1999) Molybdenum enzymes and molybdenum cofactor in plants. *Crit. Rev. Plant Sci.* **8**, 33–69
4. Romão, M.J., Archer, M., Moura, I., Moura, J.J.G., LeGall, J., Engh, R., Schneider, M., Hof, P. & Huber, R. (1995) Crystal structure of the xanthine oxidase-related aldehyde oxido-reductase from *D. gigas*. *Science* **270**, 1170–1176
5. Huber, R., Hof, P., Duarte, R.O., Moura, J.J.G., Moura, I., LeGall, J., Hille, R., Archer, M. & Romão, M. (1996) A structure-based catalytic mechanism for the xanthine oxidase family of molybdenum enzymes. *Proc. Natl. Acad. Sci. U.S.A.* **93**, 8846–8851
6. Greenwood, R.J., Wilson, G.L., Pilbrow, J.R. & Wedd, A.G. (1993) Molybdenum(V) sites in xanthine oxidase and relevant analog complexes: comparison of oxygen-17 hyperfine coupling. *J. Am. Chem. Soc.* **115**, 5385–5392
7. Howes, B.D., Bray, R.C., Richards, R.L., Turner, N.A., Bennett, B. & Lowe, D.J. (1996) Evidence favoring molybdenum-carbon bond formation in xanthine oxidase action: ^{17}O- and ^{13}C-ENDOR and kinetic studies. *Biochemistry* **35**, 1432–1443
8. Xia, M., Ilich, P., Dempski, R. & Hille, R. (1997) Recent studies of the reductive half-reaction of xanthine oxidase. *Biochem. Soc. Trans.* **25**, 768–773
9. Bray, R.C. & George, G.N. (1985) Electron paramagnetic resonance studies using pre-steady-state kinetics and substitution with stable isotopes in the mechanism of action of molybdoenzymes. *Biochem. Soc. Trans.* **13**, 560–567
10. Holm, R.H. (1990) The biologically relevant oxygen atom transfer chemistry of molybdenum: from synthetic analogue systems to enzymes. *Coord. Chem. Rev.* **100**, 183–221
11. Kisker, C., Schindelin, H., Pacheco, A., Wehbi, W.A., Garrett, R.M., Rajagopalan, K.V., Enemark, J.H. & Rees, D.C. (1997) Molecular basis of sulphite oxidase deficiency from the structure of sulphite oxidase. *Cell* **91**, 973–983
12. George, G.N., Kipke, C.A., Prince, R.C., Sunde, R.A., Enemark, J.H. & Cramer, S.P. (1989) Structure of the active site of sulphite oxidase. X-ray absorption spectroscopy of the Mo(IV), Mo(V) and Mo(VI) oxidation states. *Biochemistry* **28**, 5075–5080
13. Enemark, J.H. & Young, C.G. (1993) Bioinorganic chemistry of pterin-containing molybdenum and tungsten enzymes. *Adv. Inorg. Chem.* **40**, 1–88
14. Brody, M.S. & Hille, R. (1995) The reaction of chicken liver sulphite oxidase with dimethylsulphite. *Biochim. Biophys. Acta* **1253**, 133–135
15. Schindelin, H., Kisker, C., Hilton, J., Rajagopalan, K.V. & Rees, D.C. (1996) Crystal structure of DMSO reductase: redox-linked changes in molybdopterin coordination. *Science* **272**, 1615–1621
16. Schneider, F., Loewe, J., Huber, R., Schindelin, H. & Kisker, C. (1996) Crystal structure of dimethyl sulphoxide reductase from *Rhodobacter capsulatus* at 1.88 Å resolution. *J. Mol. Biol.* **263**, 53–63
17. McAlpine, A.S., McEwan, A.G. & Bailey, S. (1998) The high resolution structure of DMSO reductase in complex with DMSO. *J. Mol. Biol.* **275**, 613–624
18. Boyington, J.C., Gladyshev, V.N., Khangulaov, S.V., Stadtman, T.C. & Sun, P.D. (1997) Crystal structure of formate dehydrogenase H: catalysis involving Mo, molybdopterin, selenocysteine and an Fe_4S_4 cluster. *Science* **275**, 1305–1308
19. Garrett, R.M., Bellisimo, D.B. & Rajagopalan, K.V. (1995) Molecular cloning of human liver sulphite oxidase. *Biochim. Biophys. Acta* **1262**, 147–156
20. Su, W., Mertens, J.A., Kanamaru, K., Campbell, W.H. & Crawford, N.M. (1997) Analysis of wild-type and mutant nitrate reductase expressed in the methylotrophic yeast *Pichia pastoris*. *Plant Physiol.* **115**, 1135–1143

9

Coenzyme B_{12} (cobalamin)-dependent enzymes

E. Neil G. Marsh

Department of Chemistry, University of Michigan, Ann Arbor, MI 48109-1055, U.S.A.

Introduction

Vitamin B_{12} is the precursor to two B_{12} coenzymes, shown in Figure 1, which, while possessing similar chemical structures, play quite different biochemical roles. Key to the biological function of both coenzymes is their ability to form a unique, stable, covalent bond between the central cobalt atom of the coenzyme and carbon. One form, methylcobalamin (MeCbl), is involved in methylation reactions in which the methyl group is transferred to and from cobalt. The other form, adenosylcobalamin (AdoCbl), serves as a source of carbon-based free radicals that are 'unmasked' by homolysis of the bond between cobalt and the 5′ carbon of adenosine. B_{12} is found in both archaebacteria and eubacteria, as well as in eukaryotes; only plants do not appear to use B_{12}. This suggests that, despite its complex structure, B_{12} arose early in evolution and may even be of prebiotic origin [1].

Both MeCbl and AdoCbl play essential roles in the metabolism of higher eukaryotes [2]. In humans, lack of B_{12} in the diet, or an inability to absorb it, is the cause of pernicious anaemia. MeCbl is involved in the methylation of homocysteine to form methionine by methionine synthase as part of the methionine salvage pathway; homocysteine is toxic in high concentrations and may be responsible for many of the symptoms of pernicious anaemia. AdoCbl is the coenzyme for methylmalonyl-CoA mutase, an enzyme that converts methylmalonyl-CoA to succinyl-CoA, which is an essential step in the metabolism of odd-chain fatty acids.

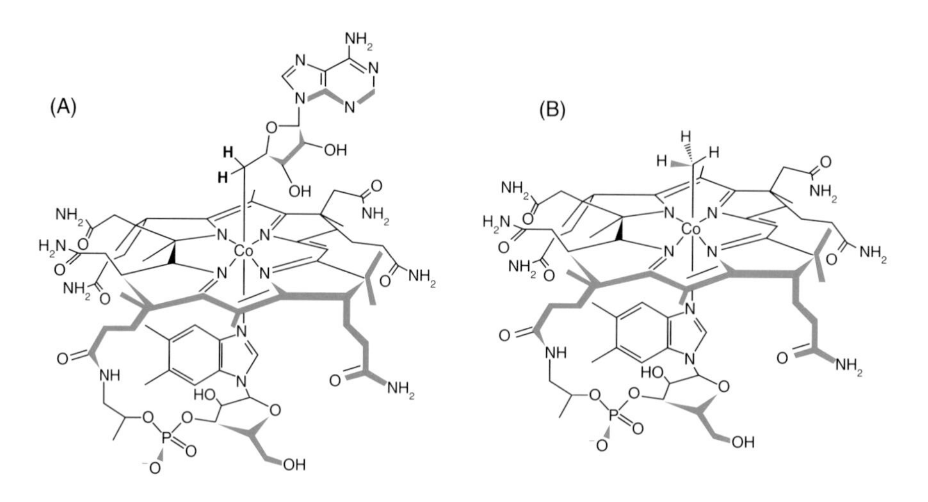

Figure 1. The structures of the B_{12} coenzymes
(**A**) AdoCbl; (**B**) MeCbl.

Microbes use B_{12} in a wide range of reactions. AdoCbl is employed in a variety of unusual isomerizations involving the cleavage of carbon–carbon, carbon–oxygen and carbon–nitrogen bonds [3]. In general, these reactions form part of degradative pathways that allow anaerobic bacteria to ferment various carbon sources. One notable exception is the reduction of ribonucleotide triphosphates to deoxyribonucleotide triphosphates that in some bacteria is catalysed by an AdoCbl-dependent ribonucleotide reductase [4]. B_{12}-like molecules (cobamides) are also involved in a variety of C-1 metabolic reactions associated with the biosynthesis of acetate and methane from CO_2 and H_2 [5].

Structures of B_{12} coenzymes

Coenzyme B_{12}, or cobalamin (Figure 1), is derived from the same porphyrin precursor as haem and chlorophyll, but has undergone further elaborate tailoring. Indeed, it has the most complex structure of any biological cofactor, requiring more than 20 genes for its biosynthesis [6]. At the heart of the coenzyme is the cobalt atom that is chelated by a macrocyclic ring called a corrin. The corrin ring is smaller than a porphyrin ring due to the loss of one carbon that used to bridge two of the pyrrole rings; this may be an adaptation for binding cobalt, which has a smaller atomic radius than iron. The macrocyclic delocalized double-bond system is extensively reduced, which results in the ring being distorted from planarity and possessing considerable flexibility. As discussed later, this flexibility may be important for its chemical function. Eight methyl groups are added to the periphery of the corrin during its synthesis; the strategic positioning of these methyl groups prevents the otherwise favourable oxidation of the double bonds.

Tethered to one of the propionamide side chains of the corrin is a nucleotide-derived 'tail', which includes a heterocyclic base that co-ordinates cobalt from below. In animals the base is always dimethylbenzimidazole; however, bacteria use a variety of bases, including adenine. In some cases the base can be altered by adding the appropriate heterocycle to the fermentation medium. The exact nature of the lower axial ligand to cobalt does not appear to be important; indeed, in some enzymes this ligand is displaced by a histidine residue from the protein upon binding to the enzyme. The function of the axial base is not fully understood. Studies on model compounds suggest that both the pK_a and the steric bulk of the axial base can influence the bond-dissociation energy of the opposing cobalt–carbon bond [7]. However, the extent to which these factors are important in enzymic catalysis is presently unclear.

Cobalt has three readily accessible oxidation states, Co(I), Co(II) and Co(III); all three are important in the functioning of the B_{12} coenzymes. In vitamin B_{12} (aquocobalamin) cobalt is in the most stable Co(III) oxidation state and, at physiological pH, the upper axial ligand is water. Aquocobalamin can undergo a two-electron reduction of cobalt to Co(I) to form cob(I)alamin, which is an extremely potent nucleophile. Cob(I)alamin is readily alkylated by a range of electrophilic reagents to form cobalt–carbon bonds in which cobalt is formally re-oxidized to the Co(III) state. In biology, methyltetrahydrofolate (Me-H_4-folate) and *S*-adenosylmethionine frequently function as methyl donors in the formation of MeCbl, and adenosine triphosphate serves as the adenosyl donor in the formation of AdoCbl. The dissociation energy of the cobalt–carbon bond is about 30 kcal/mol [8], which is very weak for a covalent bond, and it is this reactivity that Nature exploits to catalyse a variety of chemically difficult and unusual reactions. The lability of the cobalt–carbon bond also makes it sensitive to photolysis. Therefore, studies of most cobalamin-containing enzymes must be carried out in dim light.

Interaction of B_{12} with proteins

A major obstacle to a detailed understanding of B_{12}-mediated catalysis has been the lack of structural data for B_{12} enzymes. However, two crystal structures of B_{12} enzymes have been solved recently. These are the MeCbl-binding region of *Escherichia coli* methionine synthase [9] and AdoCbl-dependent methylmalonyl-CoA mutase from *Propionibacterium shermannii* [10], the structures of which are shown in Figure 2.

Sequence similarities between these two enzymes initially suggested the presence of a conserved B_{12}-binding domain. The structure of the methionine synthase fragment showed that this conserved domain comprises a five-stranded β-sheet flanked by five α-helices and is a variant of the canonical 'Rossman' nucleotide-binding fold. Most surprisingly, the structure revealed that upon binding B_{12}, the nucleotide tail of the coenzyme that co-ordinates cobalt is displaced by a histidine residue from the protein. The tail is buried between two

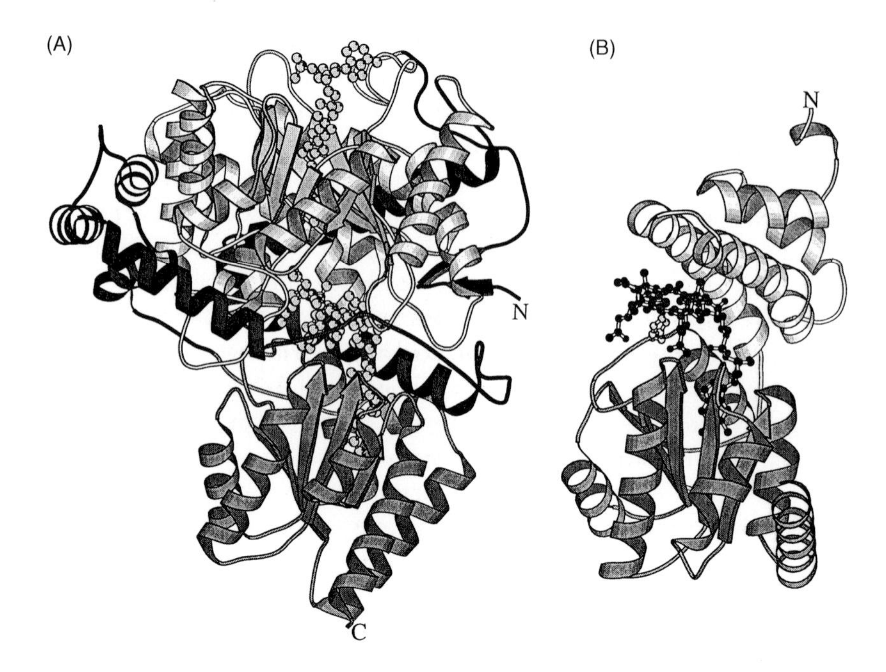

Figure 2. The structures of (A) methylmalonyl-CoA mutase (active α subunit) and (B) the B_{12}-binding fragment of methionine synthase
The cobalamin and the histidine that co-ordinates cobalt are depicted as ball-and-stick models, as is the coenzyme-A molecule bound to methylmalonyl-CoA mutase. The lower conserved B_{12}-binding domain is shown by dark shading.

helices in a groove that helps anchor the coenzyme to the protein. The histidine residue is part of a conserved Asp-X-His-X-X-Gly motif found in several cobalamin-dependent enzymes. The histidine forms a hydrogen bond with the conserved aspartate, which in turn forms a hydrogen bond with a serine residue, in a manner reminiscent of the serine proteases. Most of the other protein–coenzyme contacts occur through carbonyl groups on the backbone of the protein, which may explain the overall lack of sequence conservation in B_{12} enzymes. In methionine synthase the upper face of the coenzyme is covered by a four-helix bundle, in which a phenylalanine residue makes contact with the cobalt-bound methyl group. This residue has been shown to stabilize the cofactor by protecting the methyl group from photolysis.

The B_{12}-binding domain of methylmalonyl-CoA mutase has a very similar structure, except that a lysine replaces the serine residue found in methionine synthase as the third member of the hydrogen-bonding network. The upper face of the coenzyme makes contact with an eight-stranded β-barrel domain. One end of the barrel forms a hydrophobic chamber over the reactive face of the coenzyme, providing an inert active site in which free radicals can be contained. The substrate, methylmalonyl-CoA, enters through a narrow channel running down the centre of the barrel, effectively sealing the active site from solvent. The pantetheine chain of coenzyme A almost exactly spans the length of the β-strands of the barrel, leaving the methylmalonyl group poised above the coenzyme ready to react. Unfortunately, the adenosyl ligand to cobalt has not been resolved in the structure, so the question of how the coenzyme and substrate interact remains unanswered.

Although both of these structures show a histidine residue co-ordinated to cobalt, this is probably not generally true of B_{12} enzymes. The conserved Asp-X-His-X-X-Gly motif is only evident in the carbon skeleton mutases and some methyl transferases. EPR experiments employing ^{15}N-labelled enzymes indicate that in the cases of AdoCbl-dependent diol dehydrase and ribonucleotide reductase, the coenzyme 'tail' remains co-ordinated to cobalt when bound to the protein. Similar experiments on the corrinoid iron–sulphur protein of carbon monoxide dehydrogenase show that there is no axial nitrogenous ligand co-ordinating the cobamide [11].

AdoCbl-dependent isomerases

About 12 AdoCbl-dependent isomerases are known, which may be divided into three sub-groups dependent upon the nature of the substrate. First, there are those that catalyse the migration of hydroxy or amino groups in vicinal diols or amino alcohols, followed by dehydration or deamination to yield aldehydes. Second, there are the aminomutases, which catalyse the 1,2 migration of an amino group within an amino acid, and also require pyridoxal phosphate as an additional coenzyme. Finally, perhaps the most unusual type of isomerizations are those involving carbon-skeleton rearrangements of carboxylic or amino acids. In each case, the rearrangements involve a 1,2 interchange of a hydrogen atom on one carbon with an electron-withdrawing group, X, on an adjacent carbon [7].

The basic mechanistic scheme for these rearrangements is shown in Figure 3(A). In the first step, substrate binding initiates homolysis of AdoCbl to form cob(II)alamin and an adenosyl radical. Next, a hydrogen atom is transferred from the substrate to the adenosyl radical to give a substrate radical and 5′-deoxyadenosine. Then, in a poorly understood step, the mechanism of which may vary depending upon the enzyme, a 1,2 migration of the X group occurs to form a product radical. In the last step hydrogen is replaced to give the product and regenerate the adenosyl radical. The radical is subsequently sequestered from reacting with other molecules by recombination with cobalt to regenerate the coenzyme [12].

The fate of the migrating hydrogen has been examined for many of these enzymes using tritium as a tracer. For every case, it has been found that tritium becomes scrambled between the 5′ carbon of adenosine and the product, consistent with 5′-deoxyadenosine being the intermediate hydrogen carrier. The involvement of radical species is supported by EPR studies on many B_{12} enzymes, which show, in the presence of substrate, spectral features characteristic of both cob(II)alamin and an organic radical. [Neither cob(I)alamin nor

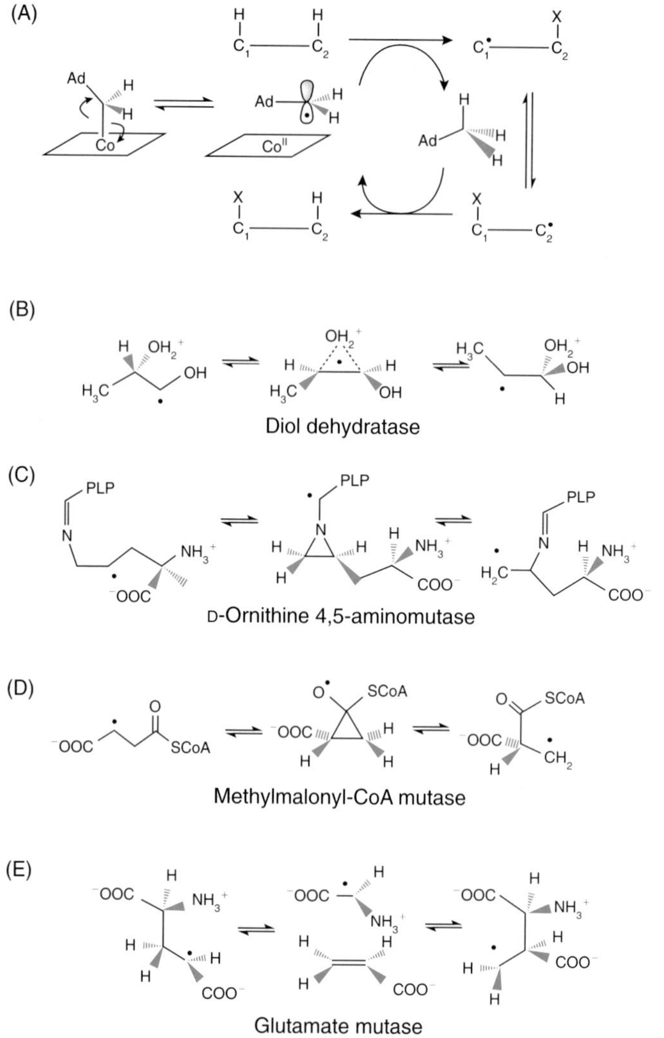

Figure 3. Mechanisms of AdoCbl-dependent isomerizations
(**A**) The minimal mechanistic scheme for the rearrangements catalysed by the AdoCbl-dependent isomerases. (**B**) The diol dehydratase-catalysed rearrangement in which migration of the hydroxy group is proposed to be facilitated by protonation. (**C**) The pyridoxal-dependent migration of an amino group catalysed by D-ornithine 4,5-aminomutase. (**D**) A mechanism for the rearrangement succinyl-CoA to methylmalonyl-CoA via an oxycyclopropyl radical intermediate. (**E**) The fragmentation–recombination mechanism for the rearrangement of L-glutamate to L-*threo*-3-methylaspartate. PLP, pyridoxal phosphate; SCoA, succinyl-CoA.

cob(III)alamin exhibit an EPR spectrum.] Furthermore, experiments in which isotopically labelled substrates are used have shown the signal from the organic radical to be modified by the presence of isotope, indicating that the radical is indeed on the substrate.

At one stage it was thought that the substrate radical would recombine with cobalt and rearrange while bound to cobalamin. This, however, is very unlikely, as the crystal structure of methylmalonyl-CoA mutase [10] shows that the substrate is too far away from the cobalt to form a substrate–cobalt bond and, indeed, 'substratyl-cobalamin' has never been detected in any AdoCbl enzyme.

Dehydratases and deaminases

This group comprises three enzymes: diol dehydrase, glycerol dehydrase and ethanolamine ammonia-lyase [13,14]. Diol dehydrase from *Klebsiella oxytoca* and glycerol dehydrase from the related bacterium *K. pneumoniae* catalyse very similar reactions, and each will use the others substrate, although less efficiently. Each comprises three subunits, α, β and γ, of molecular masses 60, 24 and 19 kDa respectively; not surprisingly, there is extensive sequence similarity between the two enzymes. Functionally, these enzymes are dimers that possess an $\alpha_2\beta_2\gamma_2$ structure and bind 2 mol of AdoCbl. Ethanolamine ammonia-lyase has been found in several bacterial species and cloned from *Salmonella typhimurium*, *Rhodococcus* sp. and *Clostridium* sp. The enzyme comprises two subunits, α (50 kDa) and β (31 kDa), assembled into a 480-kDa complex, most likely representing an $\alpha_6\beta_6$ structure. Curiously, despite its multimeric structure, the complex reportedly only binds 2 mol of AdoCbl per mol of enzyme.

Studies with ^{18}O- and deuterium-labelled (2*S*)-propan-1,2-diol have shown that the diol dehydrase reaction proceeds through the stereospecific formation of a *gem*-diol intermediate, as opposed to a radical-catalysed elimination of the hydroxy group [15]. Protonation of the hydroxy group is thought to facilitate the migration (Figure 3B). The resulting dehydration of this diol is also stereospecific and therefore must be enzyme-catalysed. The mechanism of ethanolamine ammonia-lyase is less well understood; it is not known whether a 1,1-aminoalcohol is formed as an intermediate, or if the deamination of this presumed intermediate is spontaneous or enzyme-catalysed.

Amino mutases

The amino mutases are the least extensively studied of the AdoCbl-dependent enzymes. Three examples are known, L-β-lysine[D-α-lysine]5,6-aminomutase (the same enzyme catalyses the isomerization of both D-α-lysine and L-β-lysine), D-ornithine 4,5-aminomutase and L-leucine 2,3-aminomutase [16]. The enzymes are apparently limited to clostridia (strict anaerobes), which can ferment the respective substrates. Only L-β-lysine[D-α-lysine]5,6-aminomutase and D-ornithine 4,5-aminomutase from *Clostridium sticklandii* have been studied in any detail. Both comprise several protein subunits, having molecular masses of about 200 kDa, and both require pyridoxal phosphate for activity. In addition, L-β-lysine[D-α-lysine]5,6-aminomutase requires divalent cations (Mg^{2+}) and monovalent cations (K^+ or NH_4^+) for activity, and is believed to require activation by an ATP and a pyruvate-dependent activating protein; the mechanistic significance of these other cofactors is unclear. Interestingly, there is a very similar enzyme, lysine 2,3-aminomutase, that does not use AdoCbl, but contains an iron–sulphur centre that is able to generate an adenosyl radical by a transient one-electron reduction of *S*-adenosylmethionine [17].

Pyridoxal phosphate plays an unusual role in the amino mutases by providing a low-energy pathway by which the amino group migrates from one carbon to the other. The amino group of the substrate first forms a Schiff base with pyridoxal phosphate so that the migrating nitrogen becomes sp^2 hybridized (Figure 3C). Once a radical is generated adjacent to the nitrogen-bearing carbon, the migration can proceed through a cyclic transition state in which the radical becomes delocalized on to the pyridoxal ring [17].

Carbon-skeleton mutases

Four AdoCbl-dependent enzymes are known that catalyse carbon-skeleton rearrangements: glutamate mutase, methylmalonyl-CoA mutase, α-methyleneglutarate mutase and isobutyryl-CoA mutase [18]. Glutamate mutase comprises two subunits, E (54 kDa) and S (15 kDa), the functional enzyme being an E_2S_2 tetramer containing two active sites. α-Methyleneglutarate mutase is a tetramer of subunit molecular mass 68 kDa. Methylmalonyl-CoA mutase is the only AdoCbl-dependent enzyme found in mammals as well as microbes. The mammalian enzyme is an α_2 dimer (subunit molecular mass 75 kDa), whereas the *P. shermanii* is an αβ dimer in which only the α subunit is active; the β subunit does not bind AdoCbl. With the possible exception of isobutyryl-CoA mutase, whose sequence is not yet reported, these enzymes share the conserved B_{12}-binding domain exemplified in the crystal structure of methylmalonyl-CoA mutase, including the histidine ligand to cobalt. This domain is contained within the S subunit of glutamate mutase and in the C-terminal portions of methylmalonyl-CoA and α-methyleneglutarate mutases [19].

The mechanism of these enzymes has generated much interest in the chemical community because there are no ready counterparts to these reactions in organic chemistry. Studies with model compounds designed to mimic the radical intermediates produced in the methylmalonyl-CoA mutase and α-methyleneglutarate mutase-catalysed reactions, where the migrating carbon is sp^2 hybridized, have demonstrated the feasibility of the substrate and product radicals interconverting through a cyclic transition state (Figure 3D). However, in the glutamate mutase reaction, the migrating carbon is sp^3 hybridized so that such a transition state is not possible. An alternative mechanism is hypothesized for this enzyme in which fragmentation of the glutamyl radical to give acrylate and a glycyl radical as intermediates is followed by recombination of the glycyl radical with the other end of the acrylate double bond to yield the methylaspartyl radical (Figure 3E) [18]. A similar mechanism could also be written for the other carbon-skeleton isomerases, so the actual mechanism of these rearrangements remains a matter of conjecture.

AdoCbl-dependent ribonucleotide reductase

Ribonucleotide reductase provides the only biological route to deoxyribonucleotides, the building blocks of DNA, and so is essential to all cells. In many bacteria this enzyme is AdoCbl-dependent; the most thoroughly studied such ribonucleotide reductase is from *Lactobacillus leichmannii*. It is a monomeric protein of 81.9 kDa with a sequence showing no similarities to other enzymes. The reducing power for this reaction is supplied by thioredoxin or glutaredoxin, and is shuttled to the active site via a series of redox-active disulphide bridges within the protein [20].

Although this reaction appears quite different from the isomerizations described above, AdoCbl again functions as a source of free radicals that is essential to the mechanism of reduction. Indeed, the importance of free radicals in the ribonucleotide reductase reaction is evident from the fact that there are (at least) two other classes of ribonucleotide reductase that only really differ in the cofactors used to generate free radicals [21]. For example, under anaerobic conditions *E. coli* uses an enzyme that generates adenosyl radical in a complex reaction involving a one-electron reduction of *S*-adenosylmethionine, whereas under aerobic conditions *E. coli* employs a different enzyme, in which a radical is generated by one-electron oxidation of a tyrosine residue that is stabilized by a binuclear iron cluster.

In contrast to other AdoCbl enzymes, the adenosyl radical does not react directly with the substrate, but is first transferred from adenosine to a cysteine residue at the enzyme active site (Figure 4, top scheme). The formation of this cysteinyl radical is a feature common to all ribonucleotide reductases. The cysteinyl radical removes the 3′ hydrogen of ribose so that a radical is generated *adjacent* to the carbon undergoing reduction (Figure 4, bottom scheme). The radical at the 3′ position has the effect of activating the 2′ OH to become a better leaving group. Once the hydroxy group has departed, the activated substrate radical cation is then reduced by two redox-active cysteines to give a 2′ deoxyribosyl radical, and a disulphide bridge. In the final step, the 3′ hydrogen is replaced to give the 2′ deoxyribonucleotide and regenerate the cysteinyl radical.

Interestingly, the final step in the biosynthesis of queuosine, a hypermodified base found in tRNA, has been found to be cobalamin-dependent [22]. This involves the reduction of an epoxide on a cyclopentyl ring to a double bond, a reaction clearly similar to that catalysed by ribonucleotide reductase. So far, however, nothing is known of the enzyme itself.

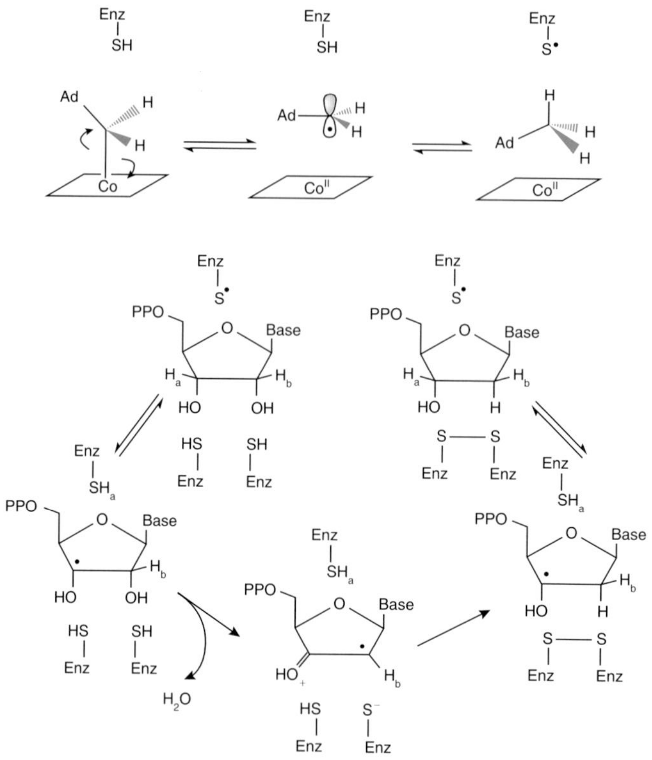

Figure 4. The mechanism of AdoCbl-dependent ribonucleotide reductase
After generating the enzyme-thiyl radical using AdoCbl (top), the reductive cycle is believed to follow essentially the same mechanism in all ribonucleotide reductases (bottom). Enz, enzyme. Subscripts a and b (H_a and H_b) denote individual hydrogen atoms.

MeCbl-dependent enzymes

Methionine synthase

Methionine synthase from *E. coli* is the most thoroughly studied MeCbl-dependent enzyme [11]. It catalyses the transfer of a methyl group from Me-H_4-folate to homocysteine (interestingly, there is also a B_{12}-independent enzyme that catalyses the same reaction). The mechanism of methionine synthase comprises two half reactions (Figure 5A). In the first, the methyl group of MeCbl is transferred to homocysteine to form methionine and the highly nucleophilic cob(I)alamin. The homocysteine is activated towards nucleophilic attack on MeCbl by co-ordination of the thiol group to a protein-bound zinc ion; this lowers the pK_a of the thiol group to favour formation of the more nucleophilic thiolate anion. In the second half reaction, MeCbl is regenerated by transfer of a methyl group from the N-5 of Me-H_4-folate to cob(I)alamin. Even though cob(I)alamin is a potent nucleophile, the methyl group of Me-H_4-folate is only weakly electrophilic, and it is likely that the

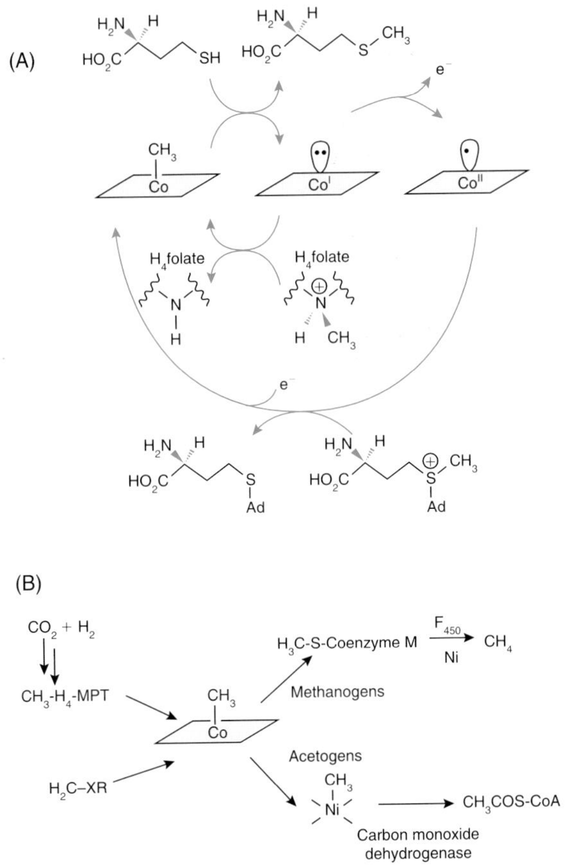

Figure 5. MeCbl-dependent methyl-transfer reactions
(**A**) Methyl-transfer reactions catalysed by methionine synthase. (**B**) The central role of cobinamide-dependent methyl-transfer reactions in the biosynthesis of methane and acetyl-CoA by methanogenic and acetogenic bacteria.

N-5 nitrogen is protonated at the active site to facilitate nucleophilic attack by cobalt.

Cob(I)alamin is a very powerful reducing agent that is even able to reduce the protons of the solvent to hydrogen gas; thus, during turnover, there is slow rate of inactivation as the coenzyme is oxidized to cob(II)alamin (Figure 5A). To overcome this, the enzyme has a reductive methylation activity that uses reduced flavodoxin, a small low-potential electron transfer protein, to reduce cobalt back to the Co(I) state. Cob(I)alamin is trapped by methylation using *S*-adenosylmethionine, which is a very good methyl donor. The energetically favourable conversion of *S*-adenosylmethionine to adenosylhomocysteine serves to offset the unfavourable reduction step.

All three methylating activities are contained within a single protein of 1227 amino acids. Proteolysis studies have shown the enzyme to have modular construction with four discrete functional domains. At the N-terminus is a

38-kDa domain that binds homocysteine; next is a 33-kDa domain that binds Me-H_4-folate. These are followed by a 27-kDa cobalamin-binding domain and, finally, a C-terminal 38-kDa domain responsible for the re-activation of the enzyme. The structures of the cobalamin-binding and activation domains have been solved separately by X-ray crystallography [9,23]. The reactions catalysed by methionine synthase pose a topographical problem, because the three methylating domains must, at various times, interact with the same face of the cobalamin cofactor. In addition, the X-ray structure shows the upper methylated face of the corrin ring to be protected by a four-helix 'cap'. It would appear, therefore, that significant changes in protein conformation must occur for all of these four domains to interact with the coenzyme. Interestingly, the co-ordination state of cobalt changes during the catalytic cycle; the histidine ligand is strongly co-ordinated to MeCbl but dissociates upon demethylation to form cob(I)alamin. It has been suggested that this may drive the required conformational changes.

Methyl transferases in methanogenesis and acetogenesis

One of the better understood systems is the *Methanobacterium thermoautotrophicum* methyltransferase, which catalyses the transfer of a methyl group from methyltetrahydromethanopterin (Me-H_4-MPT, an analogue of Me-H_4-folate) to coenzyme M (mercaptoethane sulphonate), a reaction chemically very similar to that catalysed by methionine synthase. Methyl-coenzyme M is the substrate for methyl-coenzyme M reductase that uses the nickel-containing macrocycle, coenzyme F_{450}, to reduce the methyl group to methane and regenerate coenzyme M [24].

The methyl-transferase reaction is coupled to sodium ion translocation across the cell membrane as part of the energy-transduction process in the overall reduction of CO_2 to methane. The enzyme is a multiprotein complex comprising five integral membrane proteins and three other membrane-associated cytosolic proteins [25]. One cytosolic subunit, MtrA, carries the cobamide prosthetic group (analogous to MeCbl) that serves as the intermediate methyl carrier. An integral membrane protein catalyses the first methyl transfer from Me-H_4-MPT to the cob(I)amide form of MtrA; another cytosolic protein catalyses the subsequent transfer of the methyl group to the thiol of coenzyme M. It is believed that the second methyl transfer is coupled to sodium ion translocation. EPR studies have demonstrated that the protein provides a histidine ligand to cobalt, although it does not contain the Asp-X-His-X-X-Gly motif found in methionine synthase. An interesting suggestion is that dissociation of the histidine ligand in response to the change in the oxidation state of cobalt upon demethylation may trigger a protein conformational change that could drive sodium ion translocation.

The methyl-transfer reactions involved in acetogenesis are similar to those of methanogenesis. The intermediate methyl carrier is again a discrete protein that, in addition to binding the methylcobamide, contains an iron–sulphur cluster, the mechanistic significance of which (if any) is unclear. The methyl donor is again Me-H_4-MPT, but instead of transfer to a thiol, the activated methyl group is transferred to a nickel atom at the active site of the bifunctional enzyme, carbon monoxide dehydrogenase/acetyl-CoA synthase [26]. This remarkable nickel:iron–sulphur-containing protein is able to reduce CO_2 to carbon monoxide. It then assembles a molecule of acetyl-CoA from the activated methyl group and carbon monoxide. The latter is incorporated as the carboxyl carbon of acetyl-CoA [26].

Methanogenic bacteria are also capable of using molecules such as methanol, methylamines and acetate as methyl group donors in the reductive biosynthesis of methane. Many of these processes appear to involve cobamide-dependent enzymes that presumably operate in a similar manner to the examples discussed above [5].

Perspectives

One important objective of current research in this field is to understand how the proteins modulate the reactivity of the coenzyme. In the absence of light, neither AdoCbl nor MeCbl are especially reactive in free solution under physiological conditions. Furthermore, the two coenzymes undergo quite different chemical reactions that hinge upon how the cobalt–carbon bond is broken: homolytically, to generate free radicals in the case of AdoCbl, and heterolytically, to generate a reactive nucleophile in the case of MeCbl. The recently solved structures of methionine synthase and methylmalonyl-CoA mutase have given the first insights into this problem. However, the extent to which these differences in reactivity arise through their interactions with their respective proteins, as opposed to being intrinsic properties of the coenzymes, is currently unclear.

The mechanisms by which free radicals are stabilized by the AdoCbl-dependent isomerases remain enigmatic. It has been estimated that these enzymes accelerate the rate of cobalt–carbon bond homolysis by an impressive 10^{12}-fold. Numerous model studies have been undertaken in an attempt to mimic the reactivity of the enzyme–coenzyme complex. One finding was that the cobalt–carbon bond dissociation energy is very sensitive to steric crowding at the β-face of the corrin ring. This observation led to the suggestion that these enzymes might exploit the flexibility of the corrin ring by binding the coenzyme in a conformation that introduced steric compression at the cobalt–carbon bond, thereby activating it towards homolysis. However, this hypothesis is not borne out by the structures of the two B_{12} enzymes solved so far, because in both cases, the corrin ring has very little distortion from the geometry found in free solution [9,10].

An attractive hypothesis is that enzymes may influence the reactivity of the cobalt–carbon bond through the axial nitrogenous ligand to cobalt. Early studies using alkyl cobaloximes as models showed that decreasing the basicity of the axial ligand increased the rate of cobalt–carbon bond homolysis. For this reason, the discovery of the cobalt–histidine–aspartate 'triad' in several enzymes is particularly exciting because it suggests a way in which the protein might fine tune the reactivity of the cobalt–carbon bond. Indeed, mutagenesis experiments on glutamate mutase and methionine synthase [27,28] have demonstrated the importance of these residues for enzyme activity; however, it is more difficult to determine which step(s) in the mechanism the mutations are affecting. The idea that the axial ligand may mediate redox-dependent conformational changes required in the mechanisms of some of the methyl transferases is an intriguing one that further investigations into the structure and mechanism of these enzymes will seek to prove.

Summary

- *The B_{12} or cobalamin coenzymes are complex macrocycles whose reactivity is associated with a unique cobalt–carbon bond. The two biologically active forms are MeCbl and AdoCbl and their closely related cobamide forms.*
- *MeCbl participates as the intermediate carrier of activated methyl groups. During the catalytic cycle the coenzyme shuttles between MeCbl and the highly nucleophilic cob(I)alamin form. Examples of MeCbl-dependent enzymes include methionine synthase and Me-H_4-MPT:coenzyme M methyl transferase.*
- *AdoCbl functions as a source of carbon-based free radicals that are unmasked by homolysis of the coenzyme's cobalt–carbon bond. The free radicals are subsequently used to remove non-acid hydrogen atoms from substrates to facilitate a variety of reactions involving cleavage of carbon–carbon, carbon–oxygen and carbon–nitrogen bonds. Most reactions involve 1,2 migrations of hydroxy-, amino- and carbon-containing groups, but there is also one class of ribonucleotide reductases that uses AdoCbl.*
- *The structures of two cobalamin-dependent enzymes, methionine synthase and methylmalonyl-CoA mutase, have been solved. In both cases the cobalt is co-ordinated by a histidine ligand from the protein. The significance of this binding motif is presently unclear since in other cobalamin-dependent enzymes spectroscopic evidence suggests that the coenzyme's nucleotide 'tail' remains co-ordinated to cobalt when bound to the protein.*

I am grateful to Dr. Martha Ludwig and Dr. Phil Evans for providing Figure 2. Research in my laboratory is supported in part by a grant from the National Institutes of Health, GM 55163.

References

1. Eschenmoser, A. (1988) Vitamin B_{12} – experiments concerning the origin of its molecular structure. *Angew. Chem. Int. Ed. Engl.* **27**, 5–39
2. Banerjee, R. (1997) The yin-yang of cobalamin biochemistry. *Chem. Biol.* **4**, 175–186
3. Ochiai, E.-I. (1994) Adenosylcobalamin (vitamin B_{12} coenzyme)-dependent enzymes. *Met. Ions Biol. Syst.* **30**, 255–278
4. Booker, S., Broderick, J. & Stubbe, J. (1993) Ribonucleotide reductases: radical enzymes with suicidal tendencies. *Biochem Soc. Trans.* **21**, 727–730
5. Stupperich, E. (1993) Recent advances in elucidation of biological corrinoid functions. *FEMS Microbiol. Rev.* **12**, 349–366
6. Battersby, A.R. (1994) How nature builds the pigments of life: the conquest of vitamin B_{12}. *Science* **264**, 1551–1557
7. Halpern, J. (1985) Mechanisms of coenzyme B_{12}-dependent rearrangements. *Science* **227**, 869–875
8. Finke, R.G. & Hay, B.P. (1984) Thermolysis of adenosylcobalamin: a product, kinetic, and Co-C5′ bond dissociation study. *Inorg. Chem.* **23**, 3041–3043
9. Drennan, C.L., Huang, S., Drummond, J.T., Matthews, R.G. & Ludwig, M.L. (1994) How a protein binds B_{12}: a 3.0 Å X-ray structure of B_{12}-binding domains of methionine synthase. *Science* **266**, 1669–1674
10. Mancia, F., Keep, N.H., Nakagawa, A., Leadlay, P.F., Mc Sweeney, S., Rasmussen, B., Bosecke, P., Diat, O. & Evans, P.R. (1996) How coenzyme B_{12} radicals are generated: the crystal structure of methylmalonyl-CoA mutase at 2 Å resolution. *Structure* **4**, 339–350
11. Ludwig, M.L. & Matthews, R.G. (1997) Structure-based perspectives on B_{12}-dependent enzymes. *Annu. Rev. Biochem. Mol. Biol.* **66**, 269–311
12. Marsh, E.N.G. (1995) A radical approach to enzyme catalysis. *BioEssays* **17**, 431–441
13. Toraya, T. (1994) Diol dehydrase and glycerol dehydrase, coenzyme B_{12}-dependent isozymes. *Met. Ions Biol. Syst.* **30**, 217–254
14. Faust, L.P. & Babior, B.M. (1992) Overexpression, purification, and some properties of the AdoCbl-dependent ethanolamine ammonia-lyase from *Salmonella typhimurium. Arch. Biochem. Biophys.* **294**, 50–54
15. Zagalak, B., Frey, P.A., Karabatsos, G.L. & Abeles, R.H. (1966) The stereochemistry of the conversion of D- and L-1,2-propanediols to propionaldehyde. *J. Biol. Chem.* **241**, 3028–3035
16. Baker, J.J. & Stadtman, T.C. (1982) Amino mutases. In B_{12}, vol. 2 (Dolphin, D., ed.), pp. 203–232, John Wiley & Sons, New York
17. Frey, P.A. & Reed, G.H. (1993) Lysine 2,3-aminomutase and the mechanism of the interconversion of lysine and β-lysine. *Adv. Enzymol. Rel. Areas Mol. Biol.* **66**, 1–39
18. Buckel, W. & Golding, B.T. (1996) Glutamate and 2-methyleneglutarate mutase: from microbial curiosities to paradigms for coenzyme B_{12}-dependent enzymes. *Chem. Soc. Rev.* 329–337
19. Marsh, E.N.G. & Holloway, D.E. (1992) Cloning and sequencing of glutamate mutase component S from *Clostridium tetanomorphum*. Homologies with other cobalamin-dependent enzymes. *FEBS Lett.* **310**, 167–170
20. Booker, S., Licht, S., Broderick, J. & Stubbe, J. (1994) Coenzyme B_{12}-dependent ribonucleotide reductase: evidence for the participation of five cysteine residues in ribonucleotide reduction. *Biochemistry* **33**, 12676–12685
21. Reichard, P. (1993) From RNA to DNA, why so many ribonucleotide reductases? *Science* **260**, 1773–1777
22. Frey, B., McCloskey, J., Kersten, W. & Kersten, H. (1988) New function of vitamin B_{12}: cobamide-dependent reduction of epoxyqueuosine to queuosine in tRNAs of *Escherichia coli* and *Salmonella typhimurium. J. Bacteriol.* **170**, 2078–2082
23. Dixon, M., Huang, S., Matthews, R.G. & Ludwig, M.L. (1996) Structure of the S-adenosylmethionine-binding domain of methylcobalamin-dependent methionine synthase. *Structure* **4**, 1263–1275

24. Ermler, U., Grabarse, W., Shima, S., Goubeaud, M. & Thauer, R.K. (1997) Crystal structure of methyl coenzyme M reductase: the key enzyme of biological methane formation. *Science* **278**, 1457–1462
25. Harms, U. & Thauer, R.K. (1996) The corrinoid-containing 23-kDa subunit MtrA of the energy-conserving N-5-methyltetrahydromethanopterin:coenzyme M methyltransferase complex from *Methanobacterium thermoautotrophicum*. EPR spectroscopic evidence for a histidine residue as a cobalt ligand of the cobamide. *Eur. J. Biochem.* **241**, 149–154
26. Ragsdale, S.W. & Kumar, M. (1996) Nickel-containing carbon monoxide dehydrogenase/acetyl-CoA synthase. *Chem. Soc. Rev.* **96**, 2515–2539
27. Chen, H.-P. & Marsh, E.N.G. (1997) How enzymes control the reactivity of adenosylcobalamin: effect on coenzyme binding and catalysis of mutations in the conserved histidine-aspartate pair of glutamate mutase. *Biochemistry* **36**, 7884–7889
28. Jarrett, J.T., Amaratunga, M., Drennan, C.L., Scholten, J.D., Sands, R.H., Ludwig, M.L. & Matthews, R.G. (1996) Mutations in the B_{12}-binding region of methionine synthase: how the protein controls methylcobalamin reactivity. *Biochemistry* **35**, 2464–2475

10

Oxygen reactions of the copper oxidases

James W. Whittaker

Department of Biochemistry and Molecular Biology, Oregon Graduate Institute of Science and Technology, P.O. Box 91000, Portland, OR 97291-1000, U.S.A.

Introduction

The emergence of oxygenic photosynthesis over 2.5 billion years ago led to dramatic changes in the chemistry of our planet with the introduction of a simple molecule (O_2) into the atmosphere. As a consequence of this single biological innovation, the oceans rusted, producing vast iron formations on the sea margins, and all of life was forced to adapt to a new oxidizing environment [1]. In time, molecular oxygen has become one of the central molecules of biochemistry, defining ecological boundaries, driving the energetics of respiration and participating as an oxidant in a variety of metabolic reactions, fundamentally altering the course of life in the process.

The apparent simplicity of the dioxygen molecule is deceptive, and many of its reactions are, in fact, extraordinarily complex. However, an underlying theme of oxidation–reduction chemistry can be recognized in the biochemistry of O_2. Addition of electrons (reduction) serves to activate molecular oxygen in oxygenation reactions where oxygen atoms are inserted into a molecular framework, and the unusually high affinity of O_2 for electrons allows oxygen to serve as a terminal electron acceptor in aerobic metabolism. In other reactions, the partial reduction of O_2 (by copper oxidases, for example; see below) provides a biological source of hydrogen peroxide, a versatile oxidant involved in a variety of cellular processes, including intercellular signalling and pathogenesis. The redox chemistry of oxygen is clearly the key to understanding these many and varied roles of O_2 in biochemistry.

O_2 redox chemistry

Molecular dioxygen (elemental oxygen) is highly electrophilic, being reduced by up to four electrons successively to form superoxide ($HO_2^{\cdot}$, the one-electron reduced species) and hydrogen peroxide (H_2O_2, the two-electron reduced species) as the initial reduction products [2]. Further reduction leads to formation of the hydroxyl radical and water:

$$O_2 \xrightarrow{1e^-, 1H^+} HO_2^{\cdot} \xrightarrow{1e^-, 1H^+} H_2O_2 \xrightarrow{1e^-, 1H^+} HO^{\cdot} + H_2O \xrightarrow{1e^-, 1H^+} 2H_2O \quad (1)$$

dioxygen (diradical)	superoxide (radical)	hydrogen peroxide	hydroxyl (radical)	water (hydrogen oxide)

Overall, the process is: $O_2+4H^++4e^- \rightarrow 2H_2O$. The affinity of O_2 for electrons can be measured as the oxidation–reduction potential (E) that is related to the electrochemical free energy of the reduction reaction:

$$\Delta G^\circ = -nFE^\circ \quad (2)$$

where ΔG° is the standard Gibbs free energy for a process, n is the number of electron equivalents involved in the reaction, and F is the Faraday electrochemical equivalent (9.65×10^4 Coulomb/mol). Reduction of O_2 under standard conditions (1 atm pressure, 25°C and 1 M concentration for all reactants, pH 0.0) occurs at a potential $E^\circ > 1$ V. This high potential makes O_2 an aggressive oxidant capable of attacking virtually any organic molecule, converting it to carbon dioxide and water.

The relative stabilities of the various species formed on reduction of dioxygen can be illustrated graphically in the form of a Frost (volt-equivalent) diagram for O_2 (Figure 1). In this type of plot, the stability of a given species is reflected by a lower value, and the slope for the line segment connecting any two species is proportional to the electrochemical potential for the corresponding redox couple; a more positive slope for the redox vector implies greater oxidizing power for that couple. The large positive slope for the Frost diagram between H_2O_2 and oxygen reflects the oxidizing character of this couple. Comparison with the Cu redox vector shown in the insert shows that the O_2/H_2O_2 redox couple is capable of oxidizing copper [$E_{Cu(I)/Cu(II)}=0.2$ V versus the normal hydrogen electrode].

Because protons are involved in O_2 redox chemistry (eqn. 1), the electrochemical potentials for dioxygen redox reactions are pH-dependent, decreasing with increasing pH. This is illustrated in the Frost diagram (Figure 1) by the distinct limbs for reactions of O_2 in acidic or basic conditions. Standard conditions usually cited in tables of electrochemical data (pH 0) are rather extreme and non-physiological for most organisms (the acidophilic prokaryote *Thiobacillus* and the archae *Sulfolobus* being notable exceptions); values at neutral pH (pH 7) are typically more relevant for biochemistry. The difference between the electrochemical potential for O_2 reduction under physiological

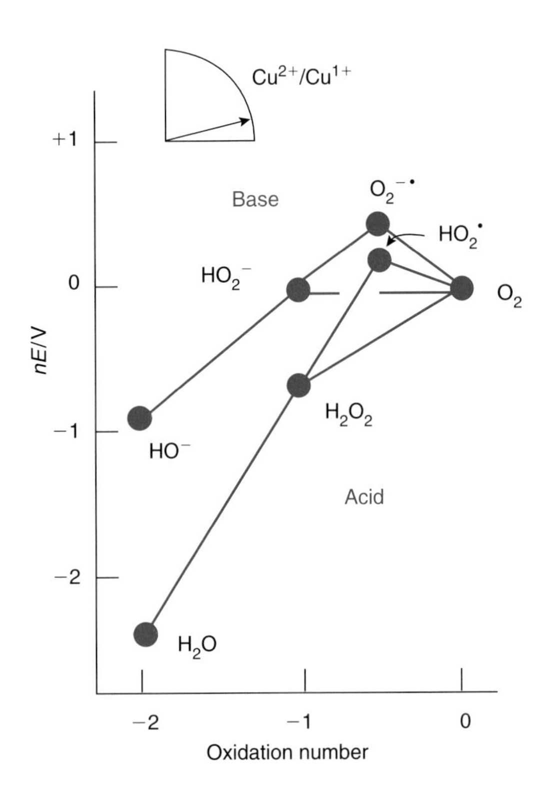

Figure 1. Frost diagram for dioxygen redox chemistry

conditions ($E^{\circ\prime}$=0.78 V) and its standard value (E°=1.23 V) emphasizes the importance of protons in dioxygen redox biochemistry. This sensitivity to pH allows the redox potential for O_2 to be tuned effectively over a significant range simply by controlling the protonation state of oxygen intermediates.

A horizontal line on the Frost diagram (Figure 1) represents an electrochemical potential of 0.0 V, corresponding to the potential of the standard hydrogen half-cell ($E^{\circ}_{H2/H+}$). Since the slopes of the O_2 diagram are strongly positive, dioxygen is thermodynamically capable of oxidizing H_2. However, even though reaction with O_2 is downhill, forming water as the stable end product, a mixture of H_2 and O_2 is quite uninteresting in the absence of a spark. Thus although O_2 is *thermodynamically* unstable in the presence of reductants, it is *kinetically* inert, with a significant activation barrier to reaction. The existence of this barrier to dioxygen reactions is the key to the enzymic control of oxygen in biochemistry.

Two major factors are responsible for the kinetic inertness of O_2: a 'spin barrier' towards non-radical reactions, and the uphill energetics of one-electron reduction. Both of these aspects of O_2 reactivity are a consequence of the electronic structure of molecular oxygen. In simplest terms, the stability of the O_2 molecule derives from the sum of the chemical bonds holding it together. A valence molecular-orbital-energy-level scheme for O_2 is shown in Figure 2. The bonding valence orbitals for molecular oxygen (3σ and 1π) are all filled, and the remaining two electrons in the ground-state configuration occupy the anti-bonding $2\pi^*$ orbital at higher energy. Since this orbital is doubly degenerate (i.e. there are two orbitals of identical energy), the lowest energy (the ground state) will be achieved by placing the electrons in different orbitals with parallel spin (↿ ↿), an arrangement that permits the electrons to effectively avoid each other and minimize electrostatic repulsion. This electronic ground state of O_2 is referred to as a triplet biradical, and has two unpaired electrons in its valence shell, contributing to a total spin of $S=1$ (*viz.* 0.5+0.5). Reaction of triplet oxygen with singlet organic molecules (spin-paired, $S=0$) to form singlet products thus violates one of the fundamental principles of physics, the conservation of angular momentum, and serves as a 'spin barrier' for reaction of O_2 with singlet molecules. This is shown by the spin inequality in eqn. (3):

$$\begin{array}{lcccc} & O_2 + XH_2(\text{red}) & \longrightarrow & 2H_2O + X(\text{ox}) \\ \text{Spin } (S) & 1 \;+\; 0 & \neq & 0 \;+\; 0 \end{array} \qquad (3)$$

The resulting kinetic inertness of O_2 mercifully spares us from spontaneous combustion. High-temperature chemistry, such as in sparks, can initiate alternative radical-chain pathways for reaction of O_2 that breach the spin barrier, conserving angular momentum. The existence of a spin barrier for dioxygen chemistry is only strictly true in bimolecular reactions, and the barrier can be circumvented by formation of covalent adducts. As we will see below, enzymes have evolved special catalytic mechanisms to overcome the spin barrier, thus allowing O_2 to participate in chemical reactions.

As O_2 is reduced, electrons enter anti-bonding orbitals, weakening the diatomic interactions and lowering the bond order, which is the sum of bond-

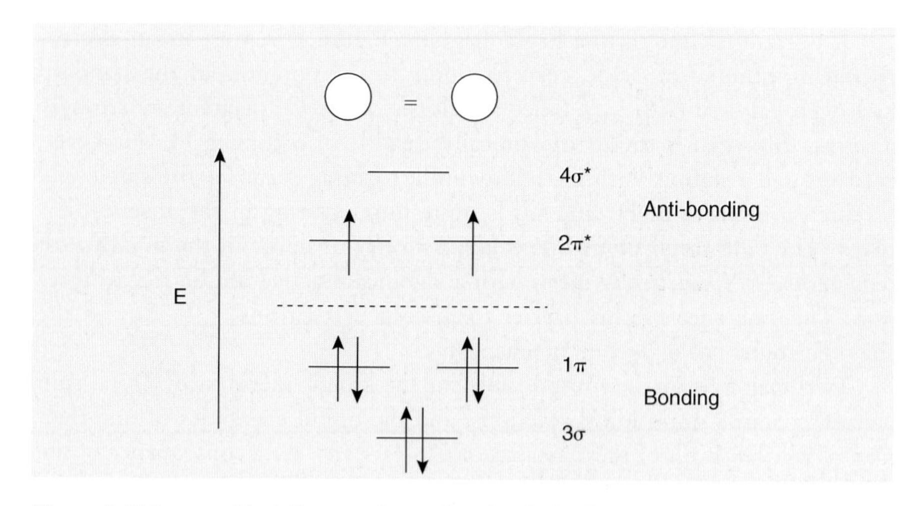

Figure 2. Valence orbital diagram for molecular O_2 in the ground state

ing and anti-bonding contributions in the valence shell. The bond order of the O_2 molecule is two, and each successive electron that is added decreases the molecule's overall stability by 0.5 of a bond. Energetically, the addition of the first electron is the most costly, as reflected in the unfavourable thermodynamics of superoxide formation (Figure 1). Superoxide is a radical, with one unpaired electron in the valence shell associated with molecular spin $S=\frac{1}{2}$ ($\underline{\uparrow\downarrow}$ $\underline{\uparrow}$). The addition of a second electron leads to the formation of the hydroperoxy dianion (not shown in Figure 2), in which the oxygen atoms are singly bonded with singlet molecular spin, $S=0$ ($\underline{\uparrow\downarrow}$ $\underline{\uparrow\downarrow}$). The third electron added to O_2 must enter the $4\sigma^*$ anti-bonding orbital, reducing the net molecular-binding energy to less than a single bond, and thus breaking the O–O bond (eqn. 1). The susceptibility of the O–O bond to reductive cleavage makes the number of electron equivalents available for reduction crucial to the mechanistic chemistry of O_2. In particular, enzymes that form H_2O_2 as one of their products must be able to constrain the delivery of electrons, controlling the redox chemistry in their active site to avoid over-reduction.

Biological chemistry of O_2

Each of the three redox forms of dioxygen (molecular oxygen, superoxide and hydrogen peroxide) has specific interactions with biological systems [2,3]. Dioxygen itself can bind reversibly with oxygen carriers, such as haemoglobin (the haemoprotein of mammalian blood), haemerythrin (a non-haem iron protein in the haemolymph of certain marine worms) or haemocyanin (a copper protein in the blood of molluscs and arthropods), forming covalent adducts with the metal cofactor in each of these proteins (see Chapter 6 in this volume). In haemoglobin, the O_2 is formally reduced by one electron to form a stable superoxo complex, while in haemerythrin and haemocyanin, molecular oxygen is formally reduced by two electrons, being bound as peroxide during transport. In each case, the reduced oxygen species remains tightly bound to the protein and is unavailable for other reactions.

In addition to these oxygen carriers, there is a wide variety of enzymes that metabolize O_2. Enzymes that catalyse the simple reduction of dioxygen, called oxidases, serve a variety of biological functions [3]. (The term 'oxidase' is used in biochemistry to generically identify enzymes using O_2 as an electron acceptor without distinguishing the number of electrons involved.) One-electron reduction of O_2 is catalysed by xanthine oxidase (a molybdopterin enzyme), which forms superoxide as a major product (see Chapter 8 in this volume). As discussed above, superoxide is a free radical, whose unpaired electron confers special reactivity that makes superoxide toxic to living cells. The toxic superoxide radical is, in fact, produced as a chemical weapon by the flavocytochrome NADPH oxidase of phagocytic leucocytes in the respiratory burst, a kind of chemical warfare of the immune system. However, most oxidases perform reductions involving an even number of electrons, thereby avoiding radical products. A large number of oxidases, including the copper oxidases that are the focus of this review, perform a two-electron reduction of O_2, forming hydrogen peroxide as the product.

Two-electron oxidases serve a variety of biological functions. As peroxide-generating systems, they may be coupled to peroxidases (haem enzymes requiring hydrogen peroxide as oxidizing substrate). Examples of this type of metabolic association include glyoxal oxidase from the wood rot fungus *Phanerochaete chrysosporium*, which is associated with lignin peroxidase and manganese peroxidase in lignin degradation, an important reaction in the global carbon cycle [4,5]. The toxicity of hydrogen peroxide is thought to be the objective of a number of oxidases, such as galactose oxidase [6–8] and glucose oxidase, which appear to have bacteriostatic roles in making the fungi that produce them more competitive in their ecological niches.

Finally, four-electron oxidases catalyse the complete reduction of O_2 to water, a reaction so energetically favourable that it can serve to drive all other metabolic processes in aerobic life. Four-electron oxidases generally have four redox centres to deliver the four reducing equivalents required in the reaction, and include the terminal oxidases (cytochrome *c* oxidase) and multicopper oxidases involved in secondary metabolism and defence in plants and fungi [ascorbate oxidase of zucchini (courgette) peeling, laccase of lacquer tree sap]. Clearly, the binding and activation of oxygen by metalloenzymes and metalloproteins is a common feature of biological systems. The mechanisms that have evolved for biological utilization of O_2 often involve redox-active metal ions (generally Fe or Cu) that donate electrons to O_2, stabilize bound intermediates and control their reactivity.

Copper oxidases

Copper oxidases (including galactose oxidase, glyoxal oxidase, amine oxidase and lysyl oxidase) are members of the two-electron oxidase family that couple oxidation of simple alcohols, aldehydes and amines to formation of hydrogen peroxide (Table 1). All known copper oxidases contain a single metal ion in a mononuclear active site in the protein. Normally, an isolated copper ion exhibits one-electron redox chemistry [Cu(I)/Cu(II)], and although higher-valency copper has been observed in some inorganic complexes, the involvement of higher oxidation states of copper in biological processes is still unconfirmed. The single-electron reactivity that is characteristic of the Cu ion is a poor match for the multi-electron redox reactions catalysed by oxidases and, in fact, all copper oxidases studied so far contain an additional redox site. This site is generally a modified protein side chain that participates in the active-site chemistry and confers two-electron reactivity (see below). The presence of these unusual catalytic cofactors has been demonstrated by spectroscopic, biochemical and crystallographic studies of the purified proteins [9–15].

Table 1. Copper oxidases

$RCH_2NH_2+O_2 \rightarrow RCHO+NH_3+H_2O_2$
Amine oxidases (TPQ)
Lysyl oxidase (LTQ)
$RCH_2OH+O_2 \rightarrow RCHO+H_2O_2$
Galactose oxidase (Tyr-Cys)
$RCHO+O_2 \rightarrow RCO_2H+H_2O_2$
Glyoxal oxidase (Tyr-Cys)

LTQ, lysine tyrosyl quinone; TPQ, topaquinone.

The presence of these unusual active-site structures in the copper oxidases raises a number of interesting questions regarding the mechanisms of this class of enzymes. What is the role of the copper? What role does the organic cofactor play? How does O_2 react? How are the cofactors formed? Each of these questions is addressed in turn below.

Role of copper

Reduced copper, Cu(I), reacts readily with dioxygen in aqueous solution [16]. Any hydrogen peroxide intermediate produced in the reaction can react with additional Cu(I), and water is generally formed as the only detectable end product of the reaction. The reactivity of Cu towards O_2 depends on the nuclearity, or number of metal ions, which determines the total number of electrons a complex can deliver, as well as the redox potential of the metal complex. The importance of nuclearity is illustrated by the binuclear Cu(I)–Cu(I) centre of the oxygen-reactive site in haemocyanin, which is capable of reducing O_2 by two electrons; this enables haemocyanin to carry oxygen as tightly bound peroxide. The redox potential of a complex determines the direction of electron transfer and basically reflects the relative stabilities of the Cu(I) and the Cu(II) species, which are determined by the particular ligand environment. Anionic ligands that donate negative charge to the metal tend to stabilize the higher oxidation state, Cu(II), whereas neutral ligands favour Cu(I). The two oxidation states are also associated with distinct ligation geometries and co-ordination numbers (the number of directly co-ordinated atoms). Low co-ordination numbers favour the lower oxidation state of a metal complex, whereas higher co-ordination numbers stabilize the higher oxidation state. Thus, Cu(I) is generally found in 2- or 3-co-ordinate (linear or T-shaped) complexes with uncharged aliphatic or aromatic amine ligands, including histidine. Cu(II) is often found in 5-co-ordinate complexes with anionic tyrosinate (phenolate) and/or cysteinate (thiolate) ligands in square-pyramidal co-ordination geometry. Enzymes that use redox-active copper must be able to accommodate both of these environments; therefore, they control the relative stability of reduced and oxidized forms of the metal to optimize the particular redox chemistry catalysed, by appropriate selection and orientation of protein side chains that serve as ligands.

The reactivity of the complex also depends on the availability of a vacant or easily exchangeable site at the metal to which O_2 can bind. In the absence of other ligands, solvent (water or hydroxide) may occupy this site. The metal (copper in this case) can enhance the reactivity of O_2 in two ways. First, by providing a one-electron pathway for the reaction, it can circumvent the spin barrier that was illustrated above in eqn. 3. Thus the reaction of triplet oxygen (S=1) with the singlet Cu(I) produces two radical products, each with a spin of $^1/_2$, rather than singlet products, and spin is conserved (eqn. 4). Secondly, covalent co-ordination to a transition-metal ion relaxes the spin restriction in the chemistry involved.

$$\begin{array}{lcccc} & O_2 + \mathrm{Cu(I)} & \longrightarrow & O_2^{-} + \mathrm{Cu(II)} & \\ \text{Spin } (S) & 1 + 0 & = & {}^1/_2 + {}^1/_2 & \end{array} \quad (4)$$

The copper environment can poise the redox potential to make the reaction with O_2 possible. However, a single Cu ion can deliver only one electron to oxygen and, as we have seen, one-electron reduction of O_2 is unfavourable. A second redox site is therefore required to provide the additional reducing equivalent. In the copper oxidases, this second redox site is a protein side chain, modified to permit oxidation–reduction reactions.

Role of the redox cofactor

Amino acids in proteins are normally non-redox active, with the notable exception of cysteine, which readily undergoes thiol–disulphide redox reactions. Although strong oxidants or ionizing radiation can generate free-radical sites in the protein by removing (or adding) one electron from (or to) the structure, free radicals derived from radiation damage in proteins tend to be very reactive and can only be detected by rapid trapping. On the other hand, certain proteins have evolved special mechanisms for stabilizing radicals that serve as reactive sites for catalysis [17–19]. Ribonucleotide reductase, one of the key enzymes in the biosynthesis of DNA, is an example of this type of free-radical enzyme. The mammalian ribonucleotide reductase contains a tyrosine free radical that stores oxidizing power in the resting enzyme [20,21]. This redox-active tyrosine owes its unusual stability to its isolation in the protein core, where it is not able to react with reductants in solution. The copper oxidases also make use of redox-active tyrosine residues for catalysis; modifications are made to the tyrosine side chain to adapt these groups to their specialized catalytic functions (Figure 3). Such amino acid modifications expand the capabilities of proteins beyond those offered by the 20-amino acid alphabet of protein structure by creating new functional groups.

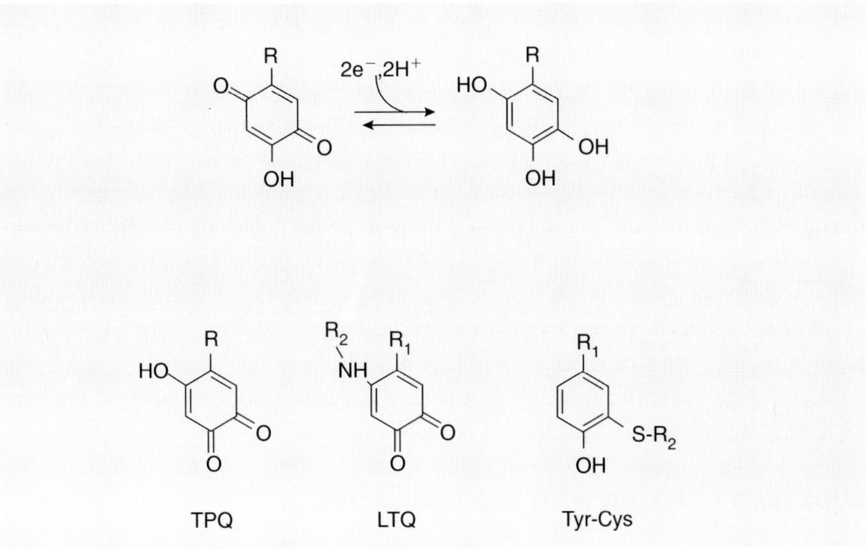

Figure 3. Redox cofactors of the copper oxidases
(Top) Redox transformation of the topaquinone (TPQ) quino-cofactor. (Bottom) Redox-active cofactors identified in copper oxidases. LTQ, lysine tyrosyl quinone.

Amine oxidases contain a single residue of topaquinone (TPQ, see Figure 3), the oxidized form of trihydroxyphenylalanine; this amino acid is derived from tyrosine by post-translational chemical modification. The 2,4,5 arrangement of the three hydroxy groups of trihydroxyphenylalanine (the reduced form of the cofactor) makes this amino acid a hydroquinone, a class of molecules that is well known to be easily oxidized by one electron to form semiquinone free radicals, and by two electrons to form quinones. This type of 'quino-cofactor' can therefore perform either one- or two-electron redox chemistry, depending on the circumstance [22]. The quinone form of this cofactor is a carbonyl compound that, in addition to redox chemistry, can react with primary amines to form an imine, also known as a Schiff base. This feature of quinone reactivity is important in amine oxidase catalysis, and reduction of the protein by the amine substrate is thought to involve formation of an imine compound. Reaction with primary amines is very characteristic of quinones and, in another copper oxidase, lysyl oxidase, the quino-cofactor is present in the protein as a lysyl imine formed by addition of a lysine residue to the oxidized side chain [23]. This is known as lysine tyrosyl quinone (Figure 3) or LTQ.

A different type of redox-active tyrosine residue is found in two other copper oxidases, galactose oxidase and glyoxal oxidase. Each of these enzymes contains a tyrosine–cysteine covalent cross-link in which the sulphur of a cysteine is attached at a position ortho to the hydroxy group of the tyrosine, as shown in Figure 3 [5,10]. Unlike the quino-cofactors described above, the tyrosine-cysteine residue pair is restricted to one-electron redox chemistry. In the active enzymes, this residue exists as a stable free radical, which is critical for catalysis, as is described below [9,24].

Why do these proteins require redox-active amino acids for catalysis? If the electron count alone were the crucial feature, a pair of metal ions (e.g. a binuclear copper complex, as found in haemocyanin) would seem to serve just as well. The special contribution of the redox-active tyrosines is the ability to donate protons as well as electrons, which are both required for the reduction of O_2. In galactose oxidase and glyoxal oxidase, the tyrosine-cysteinyl cofactor has an additional role in substrate oxidation [6,25–27]. The oxidized radical form of the cofactor (a reactive phenoxyl radical) can abstract a hydrogen atom (a proton plus an electron) from the substrate, breaking a C–H bond in the process. The cofactor thereby couples both electron and proton transfer processes by serving as a hydrogen atom-transfer agent.

Dioxygen reaction with the active site

All of these enzymes catalyse efficiently the rapid reduction of dissolved O_2 by their organic substrates. In fact, the reactions are so fast that very little is known about the individual steps, and this has led to considerable speculation about this chemistry. For each of the enzymes, the oxygen reduction reaction may be broken down into three steps: reaching the active site, reduction and elimination of the H_2O_2 product. It is thought that the paths in and out of the active site for the two dioxygen species may be distinct.

The X-ray crystal structure of the amine oxidases shows a basket-like structure with a large domain overlying the active site like a lid, isolating it from solution [12,13]. This lid probably restricts access to the active site by the amine substrates, but is not expected to be a significant barrier for O_2. Unlike large polar molecules that require an open channel to move through proteins, O_2 is sufficiently small and non-polar to diffuse through proteins with almost no impediment from the amino acid side chains; thus it can reach the active site essentially by directly diffusing into the protein core. The hydroperoxide product, on the other hand, is relatively polar, and is therefore barred from passage through the protein; it requires an exit channel from the active site. Proteins that generate peroxide at buried sites all have this type of exit channel built into their structure. In the amine oxidases, a continuous channel can be traced from the catalytic core of the protein out to the protein surface [28], and this probably represents the escape route taken by the peroxide product. These obstacles are absent in galactose oxidase, where the active site lies exposed on the protein surface [10,11], allowing the enzyme to metabolize a wide variety of primary alcohols [6].

Access to and egress from the active site are important for efficient catalysis, but are relatively uninteresting from the point of view of chemistry. The real crux of catalysis is the chemistry of the active site, i.e. the delivery of two electrons and two protons that convert O_2 into hydrogen peroxide. Where does O_2 bind? Where does it react in the active site? Because of their inherent instability, structural information for the peroxide complexes of these enzymes must be deduced from structures of the resting enzymes and their anion complexes. Peroxide bound to metal ion (Cu) in the active site will most likely co-ordinate in an easily exchangeable position, rather than displacing one of the endogenous protein ligands to the metal ion. A high-resolution crystal structure of *Escherichia coli* amine oxidase that has been covalently modified by the inhibitor 2-hydrazinopyridine shows two waters bound to adjacent sites on the metal ion [12,13] (Figure 4). Peroxide resembles a pair of cross-linked water molecules, and we might imagine that this pair of solvent molecules most closely associated with the Cu ion is a 'ghost' of the bound product, occupying the pocket designed to stabilize peroxide in the active site. The ghost structure suggests that the distal oxygen interacts with the TPQ cofactor during the reduction process, consistent with the 2-OH of TPQ being a hydrogen atom donor in the formation of a hydroperoxy anion intermediate. Dissociation of the peroxide from copper requires a second proton transfer, for which the 4-hydroxyl of the cofactor is appropriately placed.

Although a role for copper as an anchor for the reduced oxygen intermediates is appealing on the basis of general principles of metalloenzyme mechanisms, it is unclear whether the active-site metal ion in the amine oxidases

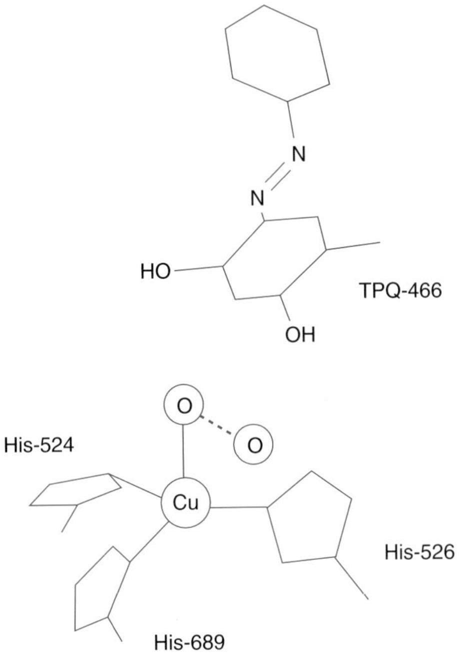

Figure 4. The structure of the active site of *E. coli* amine oxidase
The positions of two solvent molecules that may form the 'ghost' of a bound hydroperoxide are indicated. (Based on crystallographic co-ordinates, PDB ID 1SPU [12,13].)

is actually redox-active during turnover. Normally, on reduction, the organic cofactor appears to receive both electrons, forming a Cu(II)-hydroquinone derivative, in spite of the favourable environment for Cu(I) provided by the T-shaped arrangement of endogenous histidine ligands (Figure 4). However, the Cu(I)-semiquinone form is present in a temperature-dependent redox equilibrium with the Cu(II)-hydroquinone form, with the Cu(I)-semiquinone form increasing to about 40% of the active sites as the temperature is raised to the physiological range [29–32]. This form can also be stabilized by addition of cyanide, an electron-withdrawing ligand that favours formation of the reduced metal centre. Although the participation of this Cu(I)–radical complex in O_2 chemistry is still controversial, its involvement is nevertheless attractive, since it utilizes two easily oxidizable and highly reactive redox centres for reduction of O_2.

The mechanisms of the radical copper oxidases (galactose oxidase and glyoxal oxidase) are more clearly defined [6,25,26]. In each case, the active enzyme contains Cu(II) and a modified cysteinyl-tyrosine radical at the active site. Reduction of this radical by one electron produces an inactive enzyme. The crystal structure of the Cu(II)-containing inactive form of galactose oxidase [10,11] (Figure 5) shows two crystallographic water molecules in the active site, one co-ordinated to Cu in the equatorial position, and one forming a hydrogen bond to the Tyr-272 phenolate oxygen (this is part of the cysteinyl-tyrosine radical pair of the active enzyme). The position of the metal-bound water is appropriate for forming a hydrogen bond with the phenolic oxygen of

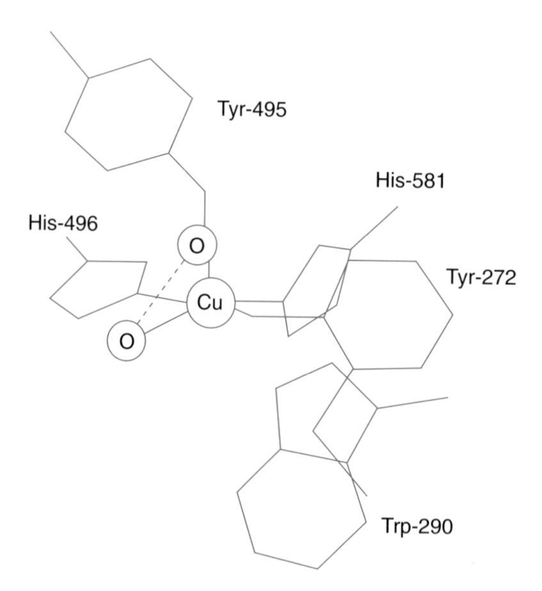

Figure 5. The structure of the active site of *Dactylium dendroides* galactose oxidase
The positions of two solvent molecules that may form the 'ghost' of a bound hydroperoxide are indicated. (Based on crystallographic co-ordinates, PDB ID 1GOG [10,11].)

Tyr-495, a residue identified as essential in catalysis [27]. This phenolate abstracts the hydroxylic proton from co-ordinated alcohol substrates, converting them to alkoxides, thus making them better reducing agents. (Figure 6 shows a postulated mechanism for the overall reaction.) A hydrogen atom is then abstracted from the adjacent methylene group by the Tyr-272 phenoxyl radical, and the resulting partially oxidized radical substrate reduces the copper to Cu(I). This results in the formation of the organic aldehyde and a 3-co-ordinate Cu(I) intermediate [33,34]. The latter reacts very rapidly with O_2 to form H_2O_2 as the end product [9]. Similar steps in reverse order are expected to occur in the reaction with oxygen, which results in the production of H_2O_2 and the regeneration of both Cu(II) and the cysteinyl-tyrosine radical. I would

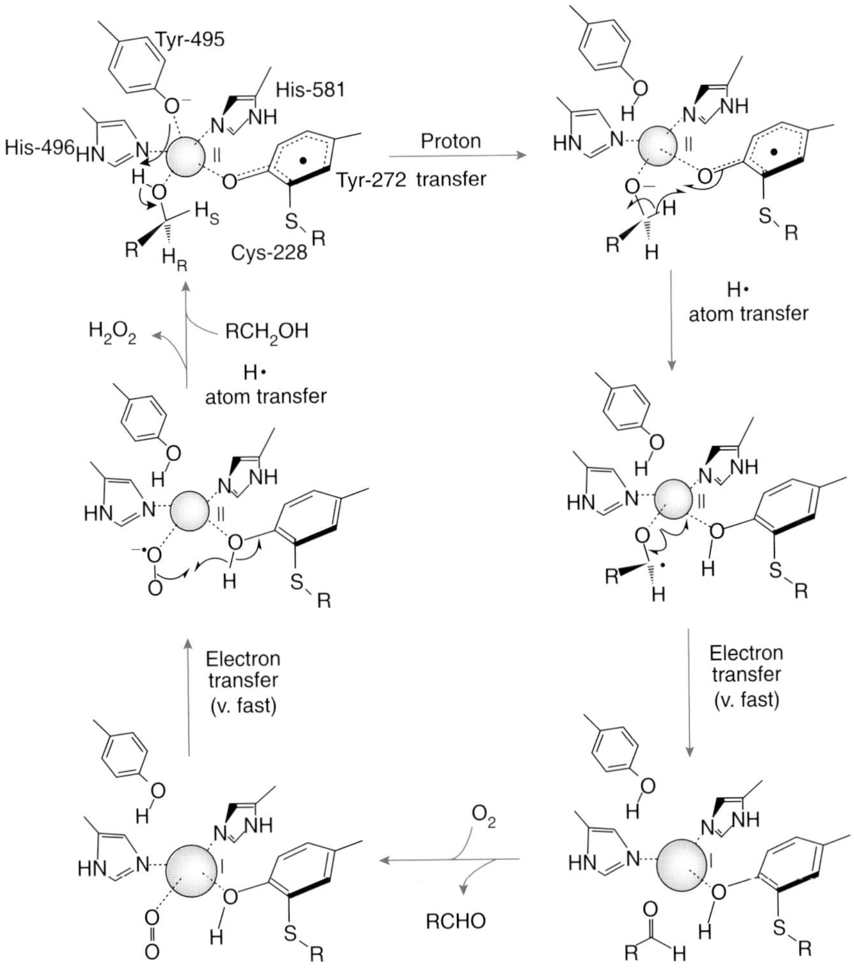

Figure 6. A postulated mechanism for galactose oxidase
See text for explanations.

therefore predict that this occurs by electron transfer to the bound O_2 from Cu(I), coupled with hydrogen atom transfer from the Tyr-272 phenol to O_2 to form the Cu(II)-phenoxyl radical pair and co-ordinated peroxide. The phenol of Tyr-495 then protonates and displaces the product as H_2O_2. This type of mechanism is ideally suited for the efficient two-electron reduction of O_2. The active site constrains both the number of electrons (two one-electron redox centres limit the number of redox equivalents available for reduction of O_2), as well as the number of protons (two protons are available, one from each of the two active-site tyrosines). This organization of the active site allows the redox reaction to occur with the lowest possible energy barriers associated with changes in buried charge. Thus changing the oxidation state of the metal by +1 and the protonation state of the ligands by -1 constitutes compensatory changes in charge at the active site, with the overall stoichiometry of $2H^+/2e^-$ (a hydrogen-molecule equivalent) transferred in the reaction. This involvement of metal ligands as acid/base catalysts illustrates a general mechanism that permits biological redox complexes to tightly couple proton and electron transfers [6,25].

Oxygen reactions in cofactor biogenesis

The modified amino acids that serve as redox cofactors in the copper oxidases are novel features of the protein structures that result from post-translational processing. In spite of the apparent complexity of the chemical changes involved, it appears that in each case the reactions occur spontaneously, the very first reaction of a pro-enzyme that converts itself it into a mature, catalytic form containing the reactive cofactor. The mechanism of cofactor biogenesis has been investigated most extensively in the amine oxidases [35–39].

Peptide sequence analysis shows that in native enzyme the cofactor exists in the position in which the gene sequence predicts there to be a tyrosine. Under anaerobic conditions, *E. coli* produces recombinant amine oxidase as a precursor protein that can be isolated and characterized [35,36], and this proenzyme has tyrosine rather than the cofactor. Addition of reduced copper and dioxygen to the purified protein results in oxidative conversion of tyrosine to quinone, a chromophoric derivative that allows absorption spectroscopy to be used to monitor the progress of the reaction. The precursor protein is colourless, whereas the mature quinoprotein is yellow ($\lambda_{max} \approx 480$ nm) as a result of electronic transitions within the cofactor. Similar results have been observed for recombinant yeast (*Hansenula*) amine oxidase expressed in *E. coli*. Site-directed mutagenesis of the tyrosine residue that is converted to TPQ in the *Arthrobacter* enzyme to a phenylalanine residue (Tyr-401→Phe) blocks this spontaneous formation of the mature cofactor [38].

Rapid-mixing stopped-flow spectroscopy has been used to measure the time course of the self-processing reaction with ms resolution [39,40]. This technique can be useful for monitoring the transient formation and decay of reaction intermediates on the basis of their unique absorption spectra. This very complex reaction, in which two oxygen atoms are added to a tyrosine residue, exhibits very simple kinetics; it is a first-order reaction requiring binding of Cu(I) to apoenzyme and subsequent reaction with O_2. Initially, it appeared that the Cu(I) oxidation state was required to form a reactive complex [39], but further studies [40] have demonstrated that Cu(II) will bind anaerobically to the apoenzyme to form a stable complex with optical and EPR spectra distinct from the TPQ-containing mature protein. This provides evidence for a significant change in structure of the metal centre in the course of the processing reaction. Interestingly, there is no evidence for the formation of a tyrosine radical in this complex, although the involvement of such a radical in the processing reaction is generally acknowledged. The elusiveness of this key radical intermediate in the cofactor biogenesis reaction is reminiscent of the behaviour of radical intermediates anticipated in a number of other biochemical reactions. For example, the catalytic free radical in ribonucleotide reductase is now known to be an active-site thiyl radical formed from a cysteine residue, rather than the Tyr-122 radical that is buried deep within the protein core, remote from the active site. Nevertheless, the Tyr-122 radical is the only radical species detected in ribonucleotide reductase, even using rapid-reaction techniques.

The complexity of the reaction is hinted at by labelling experiments that identify the origin of the quino-cofactor oxygens (water or O_2) using Raman spectroscopy [41]. Isotopic labelling changes the masses of molecular oscillators, which leads to predictable shifts in the vibrational frequencies that Raman spectroscopy detects. By choosing reaction conditions, heavy oxygen labels can be introduced either in dioxygen (as $^{18}O_2$) or in water (as $^{18}OH_2$), and the appearance of label in the quino-cofactor can be determined spectroscopically. Incubation of the mature protein with isotopically labelled water results in shifts of a single vibrational-stretch frequency, demonstrating that one quinone carbonyl oxygen (at the C-5 position of the tyrosine ring) can exchange with solvent. Reaction of apoprotein with Cu(II) in $^{18}OH_2$ and $^{16}O_2$ leads to shifts in additional bands. When the protein product is exchanged with $^{16}OH_2$, only the new bands persist, these new bands being assigned to isotope incorporated in the C-2 position. No isotopic shifts occur when the reaction is performed with $^{16}OH_2$ and $^{18}O_2$. These important experiments demonstrate that at least one of the oxygens in TPQ derives from solvent. The C-5 oxygen may derive from O_2, but, if so, it exchanges sufficiently rapidly with solvent that any isotope incorporated is washed out under the conditions of the experiment. These spectroscopic studies also provide evidence for electron delocalization between C-2 and C-4 oxygens of the quinone, and support the notion that only the C-5 oxygen has sufficient C=O carbonyl character to allow efficient formation of the imine substrate complex (Schiff base) during normal catalysis.

Analogous to the TPQ-containing amine oxidases, the Tyr-Cys-containing radical copper oxidases (galactose oxidase and glyoxal oxidase) appear to be formed from precursor proteins that self-process to generate the carbon–sulphur bond required to complete the catalytic active site. This reaction is also expected to involve radical intermediates, perhaps resulting from initial oxidation of the cysteinyl thiolate to a thiyl radical with subsequent attack on the Tyr-272. Alternatively, a tyrosine radical could be formed and react with the cysteine in a classical ortho-coupling reaction leading to an intermediate that undergoes further oxidation to complete the cofactor biogenesis.

Summary

- *The copper oxidases are a remarkable family of metalloenzymes that have evolved specialized mechanisms to accomplish the controlled reduction of dioxygen, delivering the equivalent of H_2 from organic substrates to O_2 to form hydrogen peroxide, a ubiquitous oxygen metabolite that is involved in a wide range of biological interactions.*
- *These enzymes display their virtuosity in dioxygen chemistry by harnessing the oxidizing power of that molecule not only during catalytic turnover, but also in transforming themselves in the biogenesis of their catalytic redox cofactor.*

Support for this project from the National Institutes of Health (GM 46749) is gratefully acknowledged.

References

1. Schopf, J.W. (1983) *The Earths's Earliest Biosphere*, Princeton University Press, Princeton
2. Malmstrom, B.G. (1982) Enzymology of oxygen. *Annu. Rev. Biochem.* **51**, 21–59
3. King, T.E. (1982) *Oxidases and Related Redox Systems*, Pergamon Press, Oxford
4. Kersten, P.J. (1990) Glyoxal oxidase of *Phanerochaete chrysosporium*: its characterization and activation by lignin peroxidase. *Proc. Natl. Acad. Sci. U.S.A.* **87**, 2936–2940
5. Whittaker, M.M., Kersten, P.J., Nakamura, N., Sanders-Loehr, J., Schweizer, E.S. & Whittaker, J.W. (1996) Glyoxal oxidase from *Phanerochaete chrysosporium* is a new radical-copper oxidase. *J. Biol. Chem.* **271**, 681–687
6. Whittaker, J.W. (1994) Radical copper oxidases. *Met. Ions Biol. Syst.* **30**, 315–360
7. Avigad, G., Amaral, D., Asensio, C. & Horecker, B.L. (1962) The D-galactose oxidase of *Polyporus circinatus*. *J. Biol. Chem.* **237**, 2736–2743
8. Kosman, D.J. (1985) Galactose oxidase. In *Copper Proteins and Copper Enzymes*, vol. 2 (Lontie, R., ed.), pp. 1–26, CRC Press, Boca Raton
9. Whittaker, M.M. & Whittaker, J.W. (1988) The active site of galactose oxidase. *J. Biol. Chem.* **263**, 6074–6080
10. Ito, N., Phillips, S.E.V., Stevens, C., Ogel, Z.B., McPherson, M.J., Keen, J.N., Yadav, K.D.S. & Knowles, P.F. (1991) Novel thioether bond revealed by a 1.7 Å crystal structure of galactose oxidase. *Nature (London)* **350**, 87–90

11. Ito, N., Phillips, S.E.V., Yadav, K.D.S. & Knowles, P.F. (1994) Crystal structure of a free radical enzyme, galactose oxidase. *J. Mol. Biol.* **238**, 794–814
12. Parsons, M.R., Convery, M.A., Wilmot, C.M., Yadev, K.D., Blakeley, V., Corner, A.S., Phillips, S.E.V., MacPherson, M.J. & Knowles, P.F. (1995) Crystal structure of a quinoenzyme: copper amine oxidase of *Escherichia coli* at 2 Å resolution. *Structure* **3**, 1171–1184
13. Kumar, V., Dooley, D.M., Freeman, H.C., Guss, J.M., Harvey, I., McGuirl, M.A., Wilce, M.C. & Zubak, V.M. (1996) Crystal structure of a eukaryotic (pea seedling) copper-containing amine oxidase at 2.2 Å resolution. *Structure* **4**, 943–955
14. Fontecave, M. & Eklund, H. (1995) Copper amine oxidase: a novel use for a tyrosine. *Structure* **3**, 1127–1129
15. Janes, S.M., Palcic, M.M., Scaman, C.H., Smith, A.J., Brown, D.E., Dooley, D.M., Mure, M. & Klinman, J.P. (1992) Identification of topaquinone and its consensus sequence in copper amine oxidases. *Biochemistry* **31**, 12147–12154
16. Cotton, F.A. & Wilkinson, G. (1980) *Advanced Inorganic Chemistry, a Comprehensive Text 4th edn.*, John Wiley & Sons, New York
17. Pederson, J.Z. & Finazzi-Agro, A. (1993) Protein-radical enzymes. *FEBS Lett.* **325**, 53–58
18. Klinman, J.P. (ed.) (1995) Redox-active amino acids in biology. *Methods Enzymol.* **258**
19. Sigel, H. & Sigel, A. (eds.) (1994) Metalloenzymes involving amino acid residues and related radicals. *Met. Ions Biol. Syst.* **30**
20. Sjöberg, B.-M., Reichard, P., Gräslund, A. & Ehrenberg, A. (1978) The tyrosine free radical in ribonucleotide reductase from *Escherichia coli*. *J. Biol. Chem.* **253**, 6863–6865
21. Sjöberg, B.-M. & Gräslund, A. (1983) Ribonucleotide reductase. *Adv. Inorg. Biochem.* **5**, 87–110
22. Anthony, C. (1996) Quinoprotein-catalysed reactions. *Biochem. J.* **320**, 697–711
23. Wang, S.X., Mure, M., Medzihradszky, K.F., Burlingame, A.L., Brown, D.E., Dooley, D.M., Smith, A.J., Kagan, H.M. & Klinman, J.P. (1996) A crosslinked cofactor in lysyl oxidase: redox function for amino acid side chains. *Science* **273**, 1078–1084
24. Whittaker, M.M. & Whittaker, J.W. (1990) A tyrosine-derived free radical in apogalactose oxidase. *J. Biol. Chem.* **265**, 9610–9613
25. Whittaker, J.W. (1998) Radical copper oxidases, one electron at a time. *Pure Appl. Chem.* **70**, 903–910
26. Whittaker, M.M., Ballou, D.P. & Whittaker, J.W. (1998) Kinetic isotope effects as probes of the mechanism of galactose oxidase. *Biochemistry* **37**, 8426–8436
27. Whittaker, M.M. & Whittaker, J.W. (1993) Ligand interactions with galactose oxidase: mechanistic insights. *Biophys. J.* **64**, 762–772
28. Wilce, M.C., Dooley, D.M., Freeman, H.C., Guss, J.M., Matsunami, H., McIntire, W.S., Ruggiero, C.E., Tanizawa, K. & Yamaguchi, H. (1997) Crystal structures of the copper-containing amine oxidase from *Arthrobacter globiformis* in the holo and apo forms: implications for the biogenesis of topaquinone. *Biochemistry* **36**, 16116–16133
29. Turowski, P.N., McGuirl, M.A. & Dooley, D.M. (1993) Intramolecular electron transfer rate between active-site copper and topa quinone in pea seedling amine oxidase. *J. Biol. Chem.* **268**, 17680–17682
30. Dooley, D.M. & Brown, D.E. (1996) Intramolecular electron transfer in the oxidation of amines by methylamine oxidase from *Arthrobacter* P1. *J. Biol. Inorg. Chem.* **1**, 205–209
31. Dooley, D.M., McGuirl, M.A., Brown, D.E., Turowski, P.N., McIntire, W.S. & Knowles, P.F. (1991) A Cu(I)-semiquinone state in substrate-reduced amine oxidases. *Nature (London)* **349**, 262–264
32. Steinebach, V., de Vries, S. & Duine, J.A. (1996) Intermediates in the catalytic cycle of copper-quinoprotein amine oxidase from *Escherichia coli*. *J. Biol. Chem.* **271**, 5580–5588
33. Clark, K., Penner-Hahn, J.E., Whittaker, M.M. & Whittaker, J.W. (1990) Oxidation-state assignments for galactose oxidase complexes from X-ray absorption spectroscopy. Evidence for Cu(II) in the active enzyme. *J. Am. Chem. Soc.* **112**, 6433–6434

34. Clark, K., Penner-Hahn, J.E., Whittaker, M.M. & Whittaker, J.W. (1994) Structural characterization of the copper site in galactose oxidase using X-ray absorption spectroscopy. *Biochemistry* **33**, 12553–12557
35. Tanizawa, K., Matsuzaki, R., Shimizu, E., Yorifuji, T. & Fukui, T. (1994) Cloning and sequencing of phenylethylamine oxidase from *Arthrobacter globiformis* and implication of Tyr-382 as the precursor to its covalently bound quinone cofactor. *Biochem. Biophys. Res. Commun.* **199**, 1096–1102
36. Cai, D., Williams, N.K. & Klinman, J.P. (1997) Effect of metal on 2,4,5-trihydroxyphenylalanine (topa) quinone biogenesis in the *Hansenula polymorpha* copper amine oxidase. *J. Biol. Chem.* **272**, 19277–19281
37. Matsuzaki, R., Fukui, T., Sato, H., Ozaki, Y. & Tanizawa, K. (1994) Generation of the topa quinone cofactor in bacterial monoamine oxidase by cupric ion-dependent autooxidation of a specific tyrosyl residue. *FEBS Lett.* **351**, 360–364
38. Choi, Y.H., Matsuzaki, R., Suzuki, S. & Tanizawa, K. (1996) Role of conserved Asn-Tyr-Asp-Tyr sequence in bacterial copper/2,4,5-trihydroxyphenylalanyl quinone-containing histamine oxidase. *J. Biol. Chem.* **271**, 22598–22603
39. Matsuzaki, R., Suzuki, S., Yamaguchi, K., Fukui, T. & Tanizawa, K. (1995) Spectroscopic studies on the mechanism of the topa quinone generation in bacterial monoamine oxidase. *Biochemistry* **34**, 4524–4530
40. Ruggiero, C.E., Smith, J.A., Tanizawa, K. & Dooley, D.M. (1997) Mechanistic studies of topa quinone biogenesis in phenylethylamine oxidase. *Biochemistry* **36**, 1954–1959
41. Nakamura, N., Matsuzaki, R., Choi, Y.-H., Tanizawa, K. & Sanders-Loehr, J. (1996) Biosynthesis of topa quinone cofactor in bacterial amine oxidases. Solvent origin of C-2 oxygen determined by Raman spectroscopy. *J. Biol. Chem.* **271**, 4718–4724

11

Catechol dioxygenases

Joan B. Broderick

Department of Chemistry, Michigan State University, East Lansing, MI 48824-1322, U.S.A.

Introduction

The catechol dioxygenases are a class of bacterial iron-containing enzymes that catalyse the addition of both atoms of molecular oxygen to 1,2-dihydroxybenzene (catechol) and its derivatives with subsequent cleavage of the aromatic ring [1–3]. Aromatic ring-cleaving enzymes such as the catechol dioxygenases play a central role in degradation of aromatic compounds, and thus are common in micro-organisms, particularly soil bacteria. Examples of pathways in which these enzymes are found include the chromosomally encoded pathways found in *Pseudomonas* strains for degradation of benzoate and hydroxybenzoate, called the β-ketoadipate pathway, as well as the plasmid-encoded pathway for the degradation of chlorobenzoate (Figure 1). Organisms that contain the benzoate or hydroxybenzoate degradative pathways can utilize these molecules as their sole source of carbon and energy. The plasmid-encoded haloaromatic-degrading pathways likewise enable soil bacteria to utilize halogenated organic compounds as sole sources of carbon and energy. These plasmids are, in fact, part of the machinery that allows certain bacteria not only to survive in soils polluted with halogenated organic compounds but, in doing so, to decontaminate the soils.

The catechol dioxygenases are central to the degradation of benzoate derivatives because they catalyse the critical and chemically difficult aromatic ring-cleavage reaction. The ring cleavage can occur in two different orientations relative to the vicinal diols, and this difference in cleavage site is typically used to classify the catechol dioxygenases into two groups, the intradiol- and extradiol-cleaving enzymes (Figure 2). Intradiol cleavage results in the production of *cis,cis*-muconic acids, whereas extradiol cleavage produces muconic

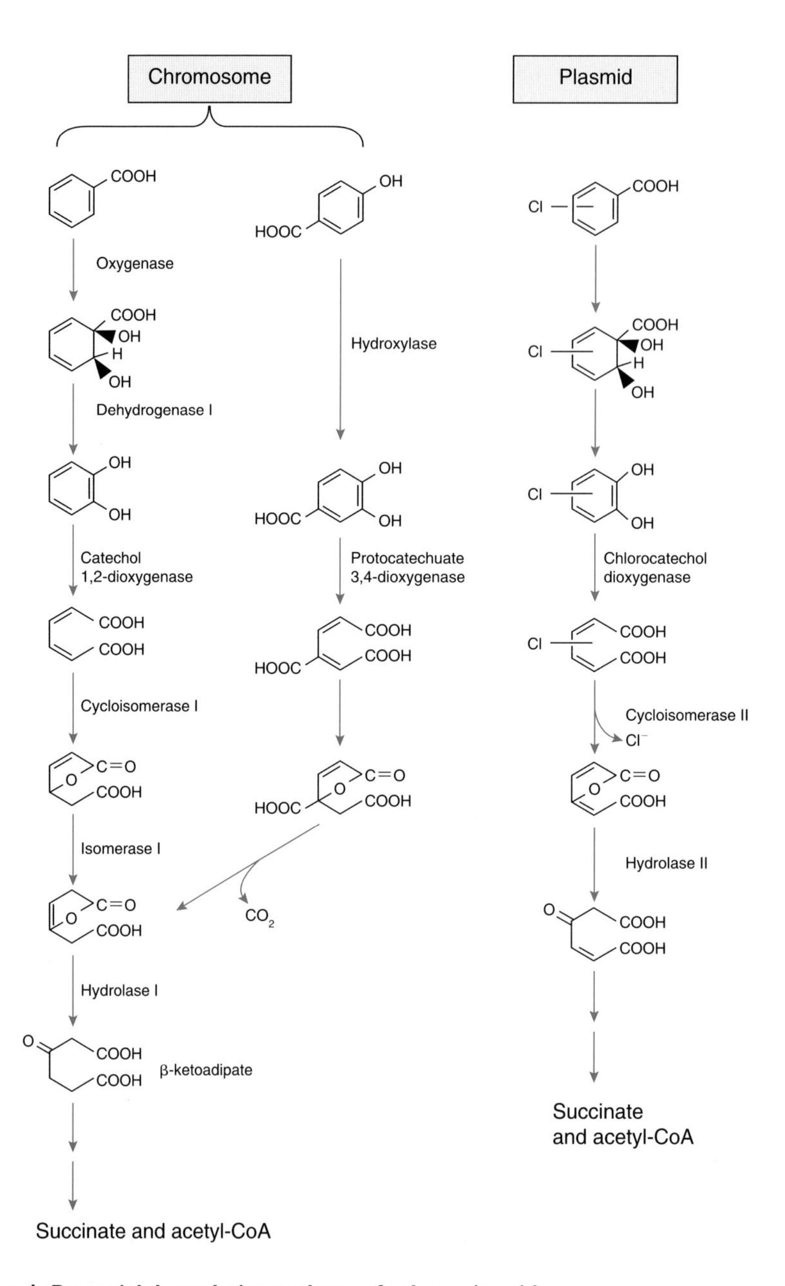

Figure 1. Bacterial degradative pathways for benzoic acids
The pathways for degradation of benzoate and *p*-hydroxybenzoate in pseudomonads are chromosomally encoded, whereas the chlorobenzoate pathway, which requires only three additional enzymes, is plasmid-encoded. Reactions in the chlorobenzoate pathway that are not labelled with enzyme names utilize the corresponding chromosomally encoded enzyme.

semialdehydes. Common among these enzymes is the absolute dependence of the activity on a bound iron atom. The intradiol catechol dioxygenases have Fe(III) at the active site, whereas the extradiol dioxygenases have Fe(II). As

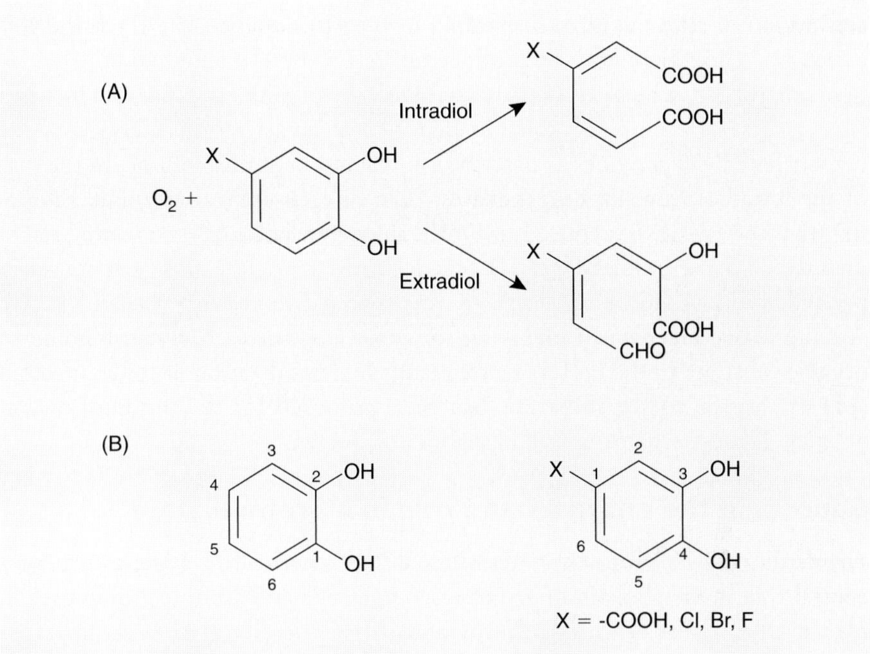

Figure 2. Modes of ring cleavage by the catechol dioxygenases
(**A**) Intradiol and extradiol cleavage. (**B**) Ring-numbering scheme for catechol and protocatechuic acid.

will be discussed below, these two categories of catechol dioxygenase exhibit fundamental differences in the iron centres, which apparently are responsible for the different regiochemistry of oxygenation.

Much interest has been focused on these enzymes from a mechanistic perspective due both to the difficulty of chemical reactions catalysed and a more general interest in biological oxygen reactions. Molecular oxygen is an essential molecule to most aerobic living systems not only due to its role as the terminal electron acceptor in electron-transport systems leading to oxidative phosphorylation, but also its critical function in the metabolism of organic compounds. A large thermodynamic driving force favours the reaction of oxygen with organic compounds; however, a kinetic barrier to oxygen reactions prevents us from spontaneously combusting in our 20% oxygen atmosphere. The kinetic barrier is a result of the difference in spin states between molecular oxygen (two unpaired electrons, triplet state) and most organic molecules (no unpaired electrons, singlet state). Thus much of the mechanistic discussion on these enzymes has focused on the question of the role of the enzyme in facilitating this 'spin-forbidden' reaction.

A metal centre such as the Fe(III) in the catechol dioxygenases could catalyse such a reaction by activating either oxygen or the organic substrate. 'Activate', in this sense, means to overcome the spin barrier to reaction. Oxygen activation can take the form of binding oxygen to a reduced Fe(II) centre followed by redox chemistry to produce an Fe(III)-bound superoxide (O_2^-) or peroxide (O_2^{2-}), which could then react with the organic substrate; such chemistry occurs, for example, in the first steps of oxygen activation catalysed by the cytochrome P450 enzymes as described in Chapter 5 of this volume. In the case of the intradiol catechol dioxygenases, such oxygen activation would require first that the Fe(III) be reduced to Fe(II), since molecular oxygen is unlikely to directly bind the oxidized Fe(III) centre. In the extradiol dioxygenases, such pre-reduction would be unnecessary, since the native enzyme contains Fe(II). An alternative mechanism involving substrate activation, in which binding of organic substrate to the Fe(III) centre results in delocalization of some unpaired spin density on to the substrate, has been proposed for the intradiol dioxygenases. These mechanisms will be elaborated below.

Sources of the enzymes and role in bioremediation

Aromatic ring-cleaving enzymes such as the catechol dioxygenases play a central role in metabolism of aromatic compounds and are thus ubiquitous in micro-organisms. Indeed, numerous intradiol-cleaving catechol 1,2-dioxygenase (CTD) and protocatechuate 3,4-dioxygenase (PCD) enzymes have been isolated from a variety of different bacteria and studied extensively since their discovery in the 1950s [1–3]. The production of catechol dioxygenases can be induced by growing certain pseudomonads on aromatic molecules as their sole carbon source. For example, benzoate and *p*-hydroxybenzoate are known to induce the production of CTD and PCD, respectively, in appropriate bacterial strains [4,5]. Likewise, *m*-toluic acid and *m*-hydroxybenzoic acid induce the production of CTD and PCD in certain *Pseudomonas* strains [6–8]. As will be discussed below, analogous halocatechol-cleaving enzymes have been isolated from *Pseudomonas* strains growing in soils contaminated with halogenated waste.

Widespread use of halogenated organic compounds in such products as pesticides, refrigerants, fire retardants, transformer-insulating materials and paints presents persistent problems in the environment, particularly because many of these materials are resistant to both chemical oxidation and biological degradation. A number of these halogenated hydrocarbons, however, have been shown to be degraded and used as sources of carbon and energy by specialized strains of aerobic and anaerobic bacteria [9–11]. These bacteria, which contain plasmids encoding the unusual degradative pathways, are typically isolated from areas contaminated with halogenated organic waste. Examples include strains that degrade pollutants such as the herbicide 2,4,5-T (2,4,5-trichlorophenoxyacetic acid), 2,6-dichlorotoluene and mono- and poly-chlorinated biphenyls, as well as other chlorinated hydrocarbons [9–11].

Certain strains of the soil bacterium *Pseudomonas* have been found that can utilize chlorobenzoate as a sole source of carbon and energy. In the case of

Pseudomonas putida, the ability to grow on chlorobenzoate is encoded on the plasmid pAC27 [12]. This plasmid contains genes for three enzymes that catalyse reactions analogous to three reactions in the chromosomally encoded degradative pathway of benzoate. As can be seen in Figure 1, the ultimate fate of the otherwise detrimental chlorobenzoate is the production of *normal* intermediates in the β-ketoadipate pathway. One of the three plasmid-encoded enzymes, chlorocatechol dioxygenase (CCD), catalyses a reaction analogous to those of CTD and PCD.

Halocarbon-degrading enzymes such as CCD are of interest due to their unique catalytic properties and their potential application in bioremediation processes. Chlorinated aromatic compounds are prominent among toxic pollutants. This, coupled with the fact that haloaromatic compounds are often notably recalcitrant to further degradation, makes the bacterial pathways for haloaromatic degradation particularly important in bioremediation processes.

Intradiol catechol dioxygenases

Spectroscopic studies: nature of the active site

Numerous spectroscopic studies on the catechol dioxygenases, particularly CTD, PCD and CCD, have provided significant insight into the iron coordination environment. All of the intradiol catechol dioxygenases studied to date contain high-spin Fe(III) in a rhombic environment and exhibit a distinct red-brown colour. The colour is a result of an electronic transition in the visible region, typically centred around 430–450 nm (Figure 3B). The wavelength and intensity of the transition ($\epsilon \approx 3000\ M^{-1} \cdot cm^{-1}$) are indicative of a charge-transfer interaction from the co-ordination of tyrosinate side chains to the Fe(III).

Resonance Raman spectroscopy has provided significant insight into the nature of this electronic transition. Resonance Raman spectroscopy is a technique that takes advantage of the occurrence of specifically enhanced intensities of vibrational modes of a chromophore that is irradiated with a laser beam of a wavelength within the absorption band of that chromophore. By specifically exciting vibrations of the chromophore molecule, scattered emission results that has frequencies characteristic of the chromophoric vibrational states. This technique, which detects that scattered light, allows one to obtain detailed vibrational information on a small chromophore embedded in a large and complex protein molecule. Resonance Raman spectroscopic studies on CTD, PCD and CCD indicate that in all cases the visible absorption band at 430–450 nm is a result of tyrosinate→Fe(III) charge-transfer transitions, indicating that tyrosine is a ligand to the iron centre. A typical resonance Raman spectrum for these enzymes is shown in Figure 3(A) [13]. The bands observed at 1183, 1266, 1507 and 1604 cm^{-1} are assigned to tyrosine-ring vibrations and are characteristic of proteins such as transferrins, which have metal–tyrosinate centres [14]. In this spectrum the C–O stretching vibration at 1266 cm^{-1} is

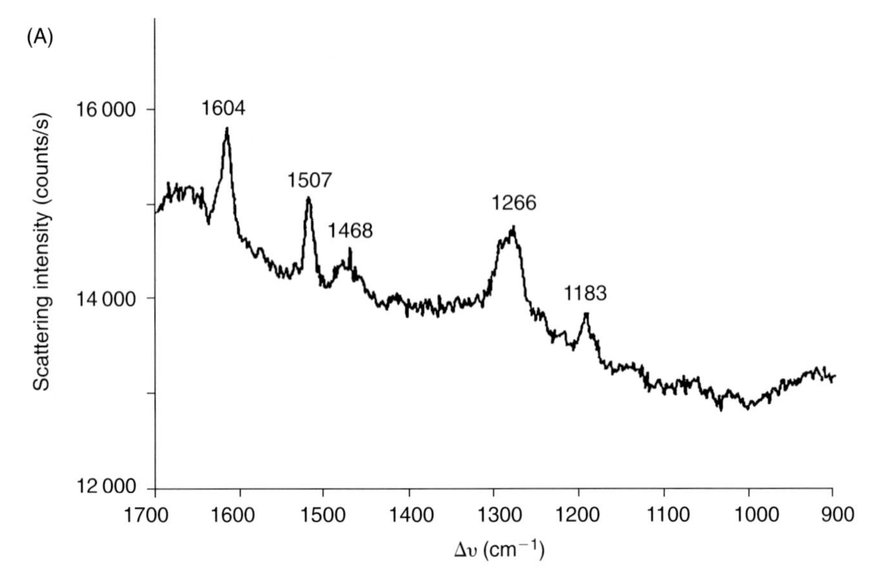

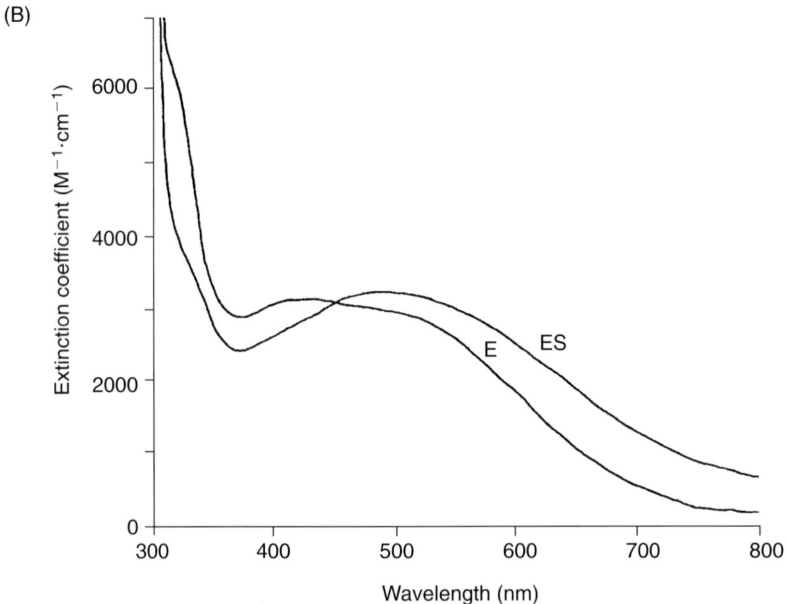

Figure 3. Resonance Raman and UV-visible spectra of CCD
(**A**) Resonance Raman spectrum of CCD. Laser excitation was at 647.1 nm, at the low-energy edge of the visible absorption band. (**B**) UV–visible absorption spectrum of CCD and the CCD–catechol complex. E, enzyme. Figures adapted from [31] with permission. © (1997) American Chemical Society.

broad; this broadness has been shown for CTD and PCD to result from the presence of two tyrosine ligands in distinct environments with slightly different C–O vibrational frequencies [15,16]. Further Raman studies on PCD, together with extended X-ray-absorption fine structure spectroscopy (EXAFS) studies, demonstrate that two histidines are co-ordinated to the iron [17]. Together, these studies suggest an active-site iron co-ordinated by four protein-derived ligands, two tyrosines in distinct environments, and two histidines. NMR spectroscopic studies on CTD corroborates this assignment of the iron ligands [18].

Binding of organic substrate to the iron centre

In order to understand the catalytic mechanism of the catechol dioxygenases, it is important to know in detail how the organic substrate binds to the enzyme. The combination of UV–visible and resonance Raman spectroscopy has shown that the catechol substrates co-ordinate directly to the Fe(III) centre [19]. It is of interest to note that the resonance Raman spectrum of the enzyme–substrate (ES) complex does not show the characteristic tyrosine vibrations, raising the possibility that binding of catecholic substrates displaces tyrosine ligands from the iron centre. However, it was proposed at the time that the lack of tyrosine vibrations in the ES complex was a result of a shift of the tyrosine→Fe(III) charge-transfer transition to higher energy, and away from the laser excitation wavelength.

Whether the substrate catechols bind to the iron through one (monodentate binding) or both (bidentate or chelated binding) of their phenolic oxygens has also been investigated. In the case of PCD, evidence points to a chelated mode of binding. For example, the EPR spectrum observed for the PCD ES complex with protocatechuate is broadened if the protocatechuate is labelled in either the 3- or 4-position with ^{17}O [20]. Magnetic nuclei such as ^{17}O will broaden EPR signals only through direct interactions, and thus these data indicate that *both* hydroxy groups of protocatechuate bind to the Fe(III).

Whittaker and Lipscomb observed broadening of the EPR signal of PCD in the presence of $H_2{}^{17}O$, demonstrating direct ligation of H_2O to the active-site iron [21]. This broadening due to $H_2{}^{17}O$ disappears when substrate is added, demonstrating that substrate and water are mutually exclusive in binding the active-site iron. EXAFS studies of the interaction of PCD with substrates indicated that although substrates apparently displace solvent and bind in a bidentate fashion, the iron co-ordination environment remains 5-co-ordinate in the ES complex [22]. Based on these results it was proposed that, upon substrate binding, not only the bound solvent, but also one of the endogenous ligands, is displaced. In contrast to the results for PCD, NMR studies of the CTD–4-methylcatechol complex indicate a monodentate substrate-binding mode for this enzyme [18].

The binding of organic substrate to the iron centre of the catechol dioxygenases is a key step in the enzymic reaction and can provide important clues as to the catalytic mechanism. As discussed above, there are two key mechanistic possibilities for the catechol dioxygenases; oxygen activation, which requires reduction of the active-site Fe(III) to Fe(II), and substrate activation, which does not require such reduction. The key question in examining the catechol dioxygenase ES complex, therefore, is whether binding of organic substrate to the iron centre results in reduction of Fe(III) to Fe(II). Several lines of evidence suggest that there is no reduction to the Fe(II) state in the anaerobic ES complexes of CTD, PCD or CCD, leaving one to conclude that catalysis proceeds via activation of substrate. If binding of organic substrate resulted in reduction of Fe(III) to Fe(II), one would expect to see a bleaching of the red colour of the enzyme and the concomitant disappearance of the visible absorption band resulting from the tyrosinate–Fe(III) interaction. However, the ES complexes remain colourful and maintain intense visible absorption bands. In fact, the ES complexes exhibit new visible absorption bands that result from catecholate→Fe(III) charge-transfer transitions.

EPR spectroscopy, which detects unpaired electrons, can also be used to look for reduction of the iron centre upon substrate binding, since Fe(III) centres typically exhibit EPR signals, whereas Fe(II) centres do not. In the case of the catechol dioxygenases, EPR spectra of the ES complexes exhibit signals characteristic of high-spin Fe(III) centres [23–25]. Mössbauer spectroscopy is another technique often used to study iron-containing proteins because it can provide specific information regarding oxidation state and other properties of the iron centre. Mössbauer studies on both CTD and PCD have shown that for both the native enzyme and the ES complexes, the iron centre is in a high-spin ferric state, with no apparent reduction to the Fe(II) state [23,24]. It is of interest to note, however, that in the CTD complex with catechol, the Mössbauer spectrum indicates some delocalization of unpaired spin density away from the ferric centre, consistent with the substrate-activation mechanism discussed above. Whereas no Mössbauer studies have been done on CCD, variable temperature/variable field magnetic susceptibility measurements, which can be used to quantify the number of unpaired electrons, indicate the presence of high-spin Fe(III) in the ES complex, consistent with the other dioxygenases [25].

Whereas no evidence of reduction of Fe(III) to Fe(II) is apparent in the static, anaerobic ES complexes, it is possible that such reduction occurs only after O_2 is added, as an intermediate in the reaction pathway. Because of the relatively intense chromophore associated with the Fe(III) centre of the catechol dioxygenases, it is possible to investigate intermediates in the catalytic mechanism using stopped-flow UV-visible spectroscopy. Such studies have been performed by Ballou and co-workers, and a number of intermediates along the reaction pathway have been identified for CTD, PCD and CCD ([26, 27], and J.B. Broderick, T.V. O'Halloran & D.P. Ballou, unpublished work). None of these species, however, is consistent with a reduced Fe(II) form of the enzymes. Thus, of the many studies that have been done to characterize the intradiol catechol dioxygenases, none has provided evidence for the reduction of the Fe(III) centre to Fe(II), which would be a required first step for an oxygen-activation mechanism.

Insights from X-ray crystallography

Detailed X-ray crystallographic studies on PCD by Lipscomb, Ohlendorf and co-workers have shed considerable light on the details of the active site of this enzyme, as well as the mode of substrate and inhibitor binding and the substrate/inhibitor-dependent changes in the active site [28–30]. As was predicted by earlier spectroscopic studies, the iron centre of PCD is 5-co-ordinate in the native state, with two endogenous (i.e. protein-derived) tyrosine ligands, two endogenous histidine ligands and one bound hydroxide co-ordinated in approximately trigonal bipyramidal geometry (Figure 4A). Note that, as predicted by resonance Raman studies (see below), the two tyrosines are in distinct environments, with Tyr-447 axial and Tyr-408 equatorial. Structures determined for PCD in complex with mono-hydroxybenzoate and 3-halo-4-hydroxybenzoate inhibitors show that these inhibitors co-ordinate to the active-site iron through the C-4 phenolate by displacing the bound hydroxide. These results were consistent with previous spectroscopic studies of the interaction of these inhibitors with PCD.

The mode of substrate binding in the intradiol catechol dioxygenases has been of much interest due to the implications for the catalytic mechanism. As was discussed above, spectroscopic studies indicate that substrates chelate the active-site iron of PCD, but that the iron remains 5-co-ordinate, presumably through loss of an endogenous iron ligand. This unusual interpretation of the spectroscopic results has now been further supported by the results of the X-ray crystal structure of the PCD–protocatechuic acid (PCA) ES complex [30]. This structure clearly shows that PCA chelates the active-site iron of PCD, displacing the endogenous Tyr-447 ligand in addition to the bound solvent (Figure 4B). This ligand displacement results in a change of the co-ordination geometry of the active-site iron from trigonal bipyramidal in the native state to square pyramidal in the ES complex. Significantly, substrate binding also appears to open up a small molecule-binding cavity in close proximity to the carbon atom of PCA that will react with O_2.

Based on their crystal structures of PCD in complex with a number of inhibitors and its native substrate, Ohlendorf and co-workers have proposed a mechanism for PCD that is consistent with the substrate activation mechanism mentioned earlier (Figure 5A). The first step in the mechanism is the binding of PCA to the Fe(III) centre concomitant with displacement of Tyr-447. The substrate then ketonizes, with a build-up of negative charge and/or radical character at C-4, which then reacts directly with O_2 to generate the peroxy intermediate (**I**; see Figure 5). Oxygen atom insertion into the C3–C4 bond produces the anhydride intermediate (**II**), which reacts with the bound O^{2-} to produce the co-ordinated muconic acid product (**III**). The muconate is then displaced by Tyr-447 and water to regenerate the active-site structure.

Dioxygenase substrates: rates and specificities

Both CTD and PCD are highly specific for their physiological substrates, catechol and PCA, respectively. With the exception of 4-fluorocatechol and 4-methylcatechol for the CTD enzymes, turnover rates for the physiological substrates of these enzymes are typically 10–1000-fold faster than the turnover

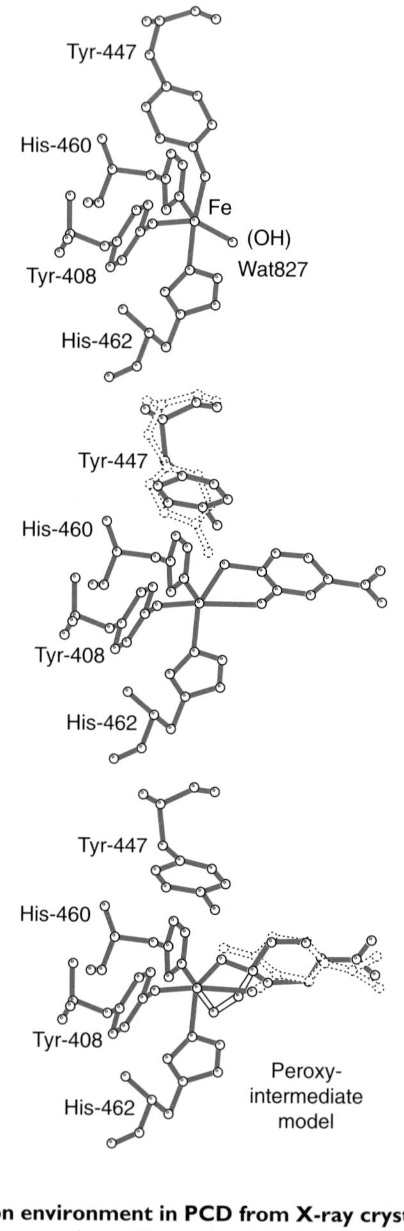

Figure 4. Iron co-ordination environment in PCD from X-ray crystallography
(**A**) Native PCD. (**B**) PCD–protocatechuate ES complex. (**C**) Hypothetical model for the peroxy intermediate. Figure was adapted with permission from [30]. © (1997) American Chemical Society. It was generously provided by Allen Orville.

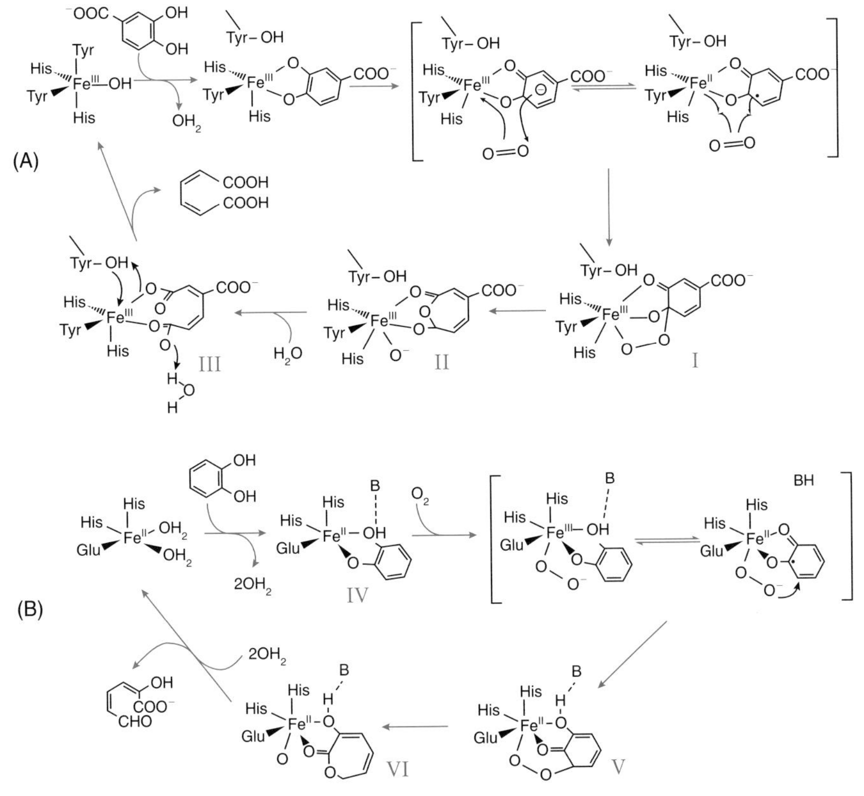

Figure 5. Proposed mechanisms for the catechol dioxygenases
(**A**) Proposed mechanisms for PCD. Adapted with permission from [30]. ©(1997) American Chemical Society. (**B**) Proposed mechanism for the extradiol catechol dioxygenases [3].

rates for most halogenated or otherwise substituted catechols and protocatechuates. In contrast, CCD exhibits broad substrate tolerance, oxygenating a number of halogenated and methylated substrates, as well as unsubstituted catechol, at significant rates (Table 1) [13]. It is of interest to note that in all of its spectroscopic properties, CCD is quite similar to CTD and PCD, indicating similar active-site iron environments, yet the catalytic properties are quite different. It is unclear at this time what chemical or physical changes have occurred in the CCD active site to yield an enzyme with such dramatically altered catalytic properties. Perhaps future crystallographic studies will shed light on this interesting question.

Extradiol catechol dioxygenases

Although extradiol cleavage is more common than intradiol cleavage in biodegradative pathways, until recently, much less was known regarding the structural and mechanistic aspects of the extradiol catechol dioxygenases due

Table 1. Comparison of normalized V_{max} values for oxygenation of substituted catechols by the catechol dioxygenases

The V_{max} values are given relative to the V_{max} of catechol for the CTD and CCD enzymes, and relative to the V_{max} of PCA for PCD. A. eutrophus, Alcaligenes eutrophus.

Catechol	*P. putida* CCD [13]	*P. putida* CTD [37]	*P. putida* PCD [38]	*P. arvilla* CTD [39]	*P. B13* CTD [40]	*P. B13* CCD [40]	*A. eutrophus* CTD [40]	*A. eutrophus* CCD [41]
Catechol	1.00	1.0	0.004	1.0	1.0	1.0	1.0	1.0
3-Chloro	1.31				0.007	1.05	0.002	1.24
4-Chloro	0.80	0.036		0.020	0.113	0.96	0.070	1.22
4-Fluoro	1.36				0.30	1.48	0.229	
4-Bromo	0.29							
3,5-Dichloro	0.83				0	0.36		1.81
4,5-Dichloro	<0.03							
Tetrachloro	<0.004				0	0	0	
3,5-Dibromo	0.44							
PCA	0.0	0	1.0					
3-Methyl	3.54	0.008	0.004	0.054	0.110	3.37	0.071	1.67
4-Methyl	2.73	0.90	0.002	0.946	0.920	3.16	0.413	
3-Methoxy	0.98	0.008			0.032	2.61		
4-Nitro	1	0						

to the relative inaccessibility of Fe(II) to most available spectroscopic techniques. Complexes of Fe(II) do not typically exhibit the kind of intense ligand→metal charge-transfer spectra responsible for the red-brown colour of the intradiol catechol dioxygenases, and therefore neither UV-visible spectroscopy nor resonance Raman spectroscopy can be used effectively to characterize the metal centre in the extradiol enzymes. In addition, Fe(II) is an integer spin system (S=2), unlike the S=5/2 Fe(III), and is therefore often EPR-silent. Despite these problems, a fair amount is now known about the active-site structure of these enzymes.

CD, magnetic CD and X-ray absorption spectroscopy have been used to determine that the Fe(II) is 5-co-ordinate in CTD, and NMR and EXAFS data provided evidence for histidine co-ordination [31–33]. The radical $^{\bullet}$NO has been found to bind to the Fe(II) centre of CTD, thereby allowing the use of EPR spectroscopy to investigate the metal centre [34]. These studies have shown that the presence of $H_2{}^{17}O$ results in EPR line broadening, suggesting the co-ordination of at least one solvent-derived ligand. In contrast to the intradiol dioxygenases, $^{\bullet}$NO can also bind to the anaerobic ES complex of the extradiol enzymes. In this case, substrate labelled with ^{17}O in either the 3- or 4-hydroxy group results in EPR line broadening; however, $H_2{}^{17}O$ does not broaden the EPR signal, indicating that chelation of the substrate to the active-site iron leads to displacement of the bound solvents. EXAFS studies on the CTD ES complex indicate an asymmetry in substrate binding that is consistent with substrate binding as a monoanion, in contrast to the intradiol enzymes for which substrate is believed to bind as a dianion [32]. This proposed difference in binding mode is consistent with the differences in iron oxidation state: Fe(III) has a higher charge and greater Lewis acidity than Fe(II), and therefore will have a greater affinity for the dianionic form of catechol.

The solution of the X-ray crystal structure of an extradiol-cleaving enzyme, 2,3-dihydroxybiphenyl 1,2-dioxygenase (BphC), confirmed the conclusions derived from the earlier spectroscopic results, and has provided further insight into the mode of substrate binding and the chemical mechanism [35,36]. The crystal structure of BphC shows a 5-co-ordinate iron centre with three endogenous protein ligands (two histidines and one glutamate), and two solvent-derived ligands (Figure 6). Thus, the nature of the ligands in the two types of enzyme (intradiol and extradiol) largely prescribes the binding of Fe(III) versus Fe(II). The intradiol enzymes have two tyrosines, which bind strongly to Fe(III), whereas the extradiol enzymes utilize carboxy and histidyl groups, which in combination bind Fe(II) better. As with the intradiol enzymes, crystallographic studies indicate that the substrate binds in a bidentate fashion to the iron centre of BphC.

A mechanism has been proposed for the extradiol catechol dioxygenases that is consistent with the available data (Figure 5B) [3]. As with the intradiol mechanism presented in Figure 5(A), the first step involves binding of the organic substrate in a chelated mode with concomitant displacement of ligands.

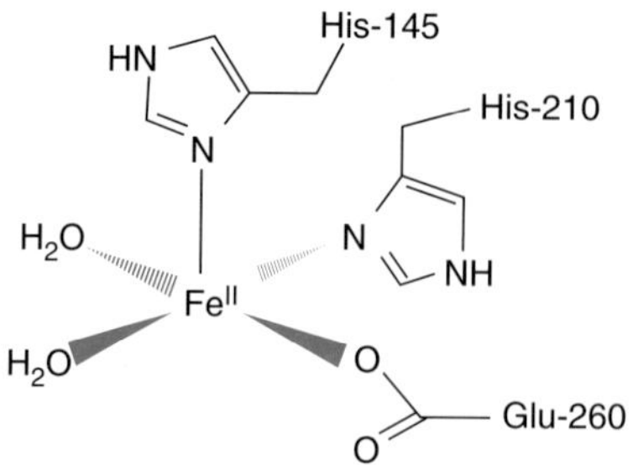

Figure 6. Iron co-ordination environment in BphC

In the case of the extradiol enzymes, two exogenous (i.e. non-protein-derived) ligands are displaced, in contrast with one exogenous and one endogenous (i.e. protein-derived) ligand for the intradiol enzymes. Presumably due to the lower charge and lower Lewis acidity of Fe(II) relative to Fe(III), catechol chelates the extradiol enzymes as a monoanion rather than a dianion (structure **IV**; see Figure 5). The next step in the mechanism is dioxygen binding, and in the case of the extradiol enzymes the presence of Fe(II) should allow direct co-ordination of O_2 to the metal centre. The observation that $^{\bullet}NO$, an O_2 analogue, can bind to the anaerobic ES complex of the extradiol enzymes provides strong support for the proposed co-ordination of O_2. Binding and reduction of O_2 leads to an Fe(III)-co-ordinated $O_2^{-\bullet}$, which can perform a nucleophilic attack on the C-3 of catechol after (or with) oxidation of the catechol by the Fe(III). The resulting intermediate (**V**) is a peroxy intermediate analogous to structure **I** (Figure 5A), but with iron in the 2^+ rather than the 3^+ oxidation state. Oxygen atom insertion into the C2–C3 bond produces intermediate **VI**, which reacts with the bound oxide to produce the product muconic semialdehyde.

Perspectives

The catechol dioxygenases are a fascinating group of enzymes due not only to their role in aromatic-ring metabolism and bioremediation, but also to the very difficult chemistry they catalyse: the controlled and specific reaction of triplet molecular oxygen with singlet organic molecules without formation of any free-radical species in solution. Through spectroscopic and X-ray crystallographic studies, much has been learned regarding the iron co-ordination environment and the mode of substrate binding for both the intradiol and extradiol enzymes. Carefully designed experiments have also provided insight into the mechanisms by which these enzymes operate. Detailed understanding of the mechanisms of these enzymes, particularly the extradiol enzymes, requires further study.

Summary

- *Catechol dioxygenases are key enzymes in the metabolism of aromatic rings by soil bacteria.*
- *Catechol dioxygenases have been found that participate in the metabolism of halogenated aromatic compounds and, in doing so, play a key role in bioremediation of halogenated pollutants.*
- *The catechol dioxygenases can be divided into two major groups: those that cleave the aromatic ring between the vicinal diols (the intradiol enzymes) and those that cleave the ring to one side of the vicinal diols (the extradiol enzymes).*
- *Whereas both types of catechol dioxygenase contain an active-site iron that is required absolutely for enzymic activity, the intradiol enzymes contain Fe(III), while the extradiol enzymes contain Fe(II). The nature of the protein ligands determines this specificity.*
- *The differences in oxidation state of the active-site iron appear to result in mechanistic differences that lead to the differing regioselectivity of the two groups of catechol dioxygenase. Mechanistic proposals based on available evidence suggest a substrate-activation mechanism for the intradiol enzymes and an oxygen-activation mechanism for the extradiol enzymes.*

References

1. Lipscomb, J.D. & Orville, A.M. (1992) Mechanistic aspects of dihydroxybenzoate dioxygenases. *Met. Ions Biol. Syst.* **28**, 243–298
2. Que, Jr., L. (1989) Oxygen activation at nonheme iron centres. In *Bioinorganic Catalysis* (Reedijk, J., ed.), pp. 347–393, Marcel Dekker, New York
3. Que, Jr., L. & Ho, R.Y.N. (1996) Dioxygen activation by enzymes with mononuclear non-heme iron active sites. *Chem. Rev.* **96**, 2607–2624
4. Feist, C.F. & Hegeman, G.D. (1969) Regulation of the *meta* cleavage pathway for benzoate oxidation by *Pseudomonas putida*. *J. Bacteriol.* **100**, 1121–1123
5. Ornston, L.N. & Stanier, R.Y. (1966) The conversion of catechol and protocatechuate to β-ketoadipate by *Pseudomonas putida*. *J. Biol. Chem.* **241**, 3776–3786
6. Nakai, C., Hori, K., Kagamiyama, H., Nakazawa, T. & Nozaki, M. (1983) Purification, subunit structure, and partial amino acid sequence of metapyrocatechase. *J. Biol. Chem.* **258**, 2916–2922
7. Nakai, C., Kagamiyama, H., Nozaki, M., Nakazawa, T., Inouye, S., Evina, Y. & Nakazawa, A. (1983) Complete nucleotide sequence of the metapyrocatechase gene on the tol plasmid of *Pseudomonas putida* mt-2. *J. Biol. Chem.* **258**, 2923–2928
8. Arciero, D.M., Lipscomb, J.D., Huynh, B.H., Kent, T.A. & Münck, E. (1983) EPR and Mössbauer studies of protocatechuate 4,5-dioxygenase: characterization of a new Fe(II) environment. *J. Biol. Chem.* **258**, 14981–14991
9. Ghosal, D., You, I.-S., Chatterjee, D.K. & Chakrabarty, A.M. (1985) Microbial degradation of halogenated compounds. *Science* **228**, 135–142
10. Reineke, W. & Knackmuss, H.-J. (1988) Microbial degradation of haloaromatics. *Ann. Rev. Microbiol.* **42**, 263–287
11. Quensen, III, J.F., Tiedje, J.M. & Boyd, S.A. (1988) Reductive dechlorination of polychlorinated biphenyls by anaerobic microorganisms from sediments. *Science* **242**, 752–754

12. Chatterjee, D.K., Kellogg, S.T., Hamada, S. & Chakrabarty, A.M. (1981) Plasmid specifying total degradation of 3-chlorobenzoate by a modified *ortho* pathway. *J. Bacteriol.* **146**, 639–646
13. Broderick, J. & O'Halloran, T.V. (1991) Overproduction, purification, and characterization of chlorocatechol dioxygenase, a non-heme iron dioxygenase with broad substrate tolerance. *Biochemistry* **30**, 7349–7358
14. Tatsuno, Y. & Saeki, Y. (1978) Resonance Raman spectra of protocatechuate 3,4-dioxygenase. Evidence for coordination of tyrosine residue to ferric iron. *J. Am. Chem. Soc.* **100**, 4614–4615
15. Que, Jr., L., Heistand, R.H. II, Mayer, R., & Roe, A.L. (1980) Resonance Raman studies of pyrocatechase-inhibitor complexes. *Biochemistry* **19**, 2588–2593
16. Que, Jr., L. & Epstein, R.M. (1981) Resonance Raman studies on protocatechuate 3,4-dioxygenase-inhibitor complexes. *Biochemistry* **20**, 2545–2549
17. Felton, R.H., Barrow, W.L., May, S.W., Sowell, A.L., Goel, S., Bunker, G. & Stern, E.A. (1982) EXAFS and Raman evidence for histidine binding at the active site of protocatechuate 3,4-dioxygenase. *J. Am. Chem. Soc.* **104**, 6132–6134
18. Que, Jr., L., Lauffer, R.B., Lynch, J.B., Murch, B.P. & Pyrz, J.W. (1987) Elucidation of the coordination chemistry of the enzyme-substrate complex of catechol 1,2-dioxygenase by NMR spectroscopy. *J. Am. Chem. Soc.* **109**, 5381–5385
19. Que, Jr., L. & Heistand, II, R.H. (1979) Resonance Raman studies on pyrocatechase. *J. Am. Chem. Soc.* **101**, 2219–2221
20. Orville, A.M. & Lipscomb, J.D. (1989) Binding of isotopically labelled substrates, inhibitors, and cyanide by protocatechuate 3,4-dioxygenase. *J. Biol. Chem.* **264**, 8791–8801
21. Whittaker, J.W. & Lipscomb, J.D. (1984) ^{17}O-Water and cyanide ligation by the active site iron of protocatechuate 3,4-dioxygenase: evidence for displaceable ligands in the native enzyme and in complexes with inhibitors or transition state analogs. *J. Biol. Chem.* **259**, 4487–4495
22. True, A.E., Orville, A.M., Pearce, L.L., Lipscomb, J.D. & Que, Jr., L. (1990) An EXAFS study of the interaction of substrate with the ferric active site of protocatechuate 3,4-dioxygenase. *Biochemistry* **29**, 10847–10854
23. Whittaker, J.W., Lipscomb, J.D., Kent, T.A. & Münck, E. (1984) *Brevibacterium fuscum* protocatechuate 3,4-dioxygenase: purification, crystallization, and characterization. *J. Biol. Chem.* **259**, 4466–4475
24. Kent, T.A., Münck, E., Pyrz, J.W. & Que, Jr., L. (1987) Mössbauer and EPR spectroscopy of catechol 1,2-dioxygenase. *Inorg. Chem.* **26**, 1402–1408
25. Broderick, J.B. (1992) Ph.D. dissertation, Northwestern University
26. Bull, C., Ballou, D.P. & Otsuka, S. (1981) The reaction of oxygen with protocatechuate 3,4-dioxygenase from *Pseudomonas putida*. *J. Biol. Chem.* **256**, 12681–12686
27. Walsh, T.A., Ballou, D.P., Mayer, R. & Que, Jr., L. (1983) Rapid reaction studies on the oxygenation reactions of catechol dioxygenase. *J. Biol. Chem.* **258**, 14422–14427
28. Ohlendorf, D.H., Lipscomb, J.D. & Weber, P.C. (1988) Structure and assembly of protocatechuate 3,4-dioxygenase. *Nature (London)* **336**, 403–405
29. Orville, A.M., Elango, N., Lipscomb, J.D. & Ohlendorf, D.H. (1997) Structures of competitive inhibitor complexes of protocatechuate 3,4-dioxygenase: multiple exogenous ligand binding orientations within the active site. *Biochemistry* **36**, 10039–10051
30. Orville, A.M., Lipscomb, J.D. & Ohlendorf, D.H. (1997) Crystal structures of substrate and substrate analog complexes of protocatechuate 3,4-dioxygenase: endogenous Fe^{3+} ligand displacement in response to substrate binding. *Biochemistry* **36**, 10052–10066
31. Mabrouk, P.A., Orville, A.M., Lipscomb, J.D. & Solomon, E.I. (1991) Variable-temperature variable-field magnetic circular dichroism studies of the Fe(II) active site in metapyrocatechase: implications for the molecular mechanism of extradiol dioxygenases. *J. Am. Chem. Soc.* **113**, 4053–4061
32. Shu, L., Chiou, Y.-M., Orville, A.M., Miller, M.A., Lipscomb, J.D. & Que, Jr., L. (1995) X-ray absorption spectroscopic studies of the Fe(II) active site of catechol 2,3-dioxygenase. Implications for the extradiol cleavage mechanism.*Biochemistry* **34**, 6649–6659

33. Bertini, I., Capozzi, F., Dikiy, A., Happe, B., Luchinat, C. & Timmis, K.N. (1995) Evidence of histidine coordination to the catalytic ferrous ion in the ring-cleaving 2,2′,3-trihydroxybiphenyl dioxygenase from the dibenzofuran-degrading bacterium *Sphingomonas* sp. strain RW1. *Biochem. Biophys. Res. Commun.* **215**, 855–860
34. Arciero, D.M., Orville, A.M. & Lipscomb, J.D. (1985) [^{17}O]Water and nitric oxide binding by protocatechuate 4,5-dioxygenase and catechol 2,3-dioxygenase: evidence for binding of exogenous ligands to the active site Fe(II) of extradiol dioxygenases. *J. Biol. Chem.* **260**, 14035–14044
35. Han, S., Eltis, L.D., Timmis, K.N., Muchmore, S.W. & Bolin, J.T. (1995). Crystal structure of the biphenyl-cleaving extradiol dioxygenase from a PCB-degrading pseudomonad. *Science* **270**, 976–980
36. Senda, T., Sugiyama, K., Narita, H., Yamamoto, T., Kimbara, K., Fukuda, M., Sato, M., Yano, K. & Mitsui, Y. (1996) Three-dimensional structures of free form and two substrate complexes of an extradiol ring-cleavage type dioxygenase, the BphC enzyme from *Pseudomonas* sp. strain KKS102. *J. Mol. Biol.* **255**, 735–752
37. Nakai, C.T., Nakazawa, T. & Nozaki, M. (1988) Purification and preperties of catechol 1,2-dioxygenase from *Pseudomonas putida* mt-2 in comparison with that from *Pseudomonas arvilla* C-1. *Arch. Biochem. Biophys.* **267**, 701–713
38. Fujisawa, H. & Hayaishi, O. (1968) Protocatechuate 3,4-dioxygenase. I. Crystallization and characterization. *J. Biol. Chem.* **243**, 2673–2681
39. Fujiwara, M., Golovleva, L.A., Saeki, Y., Nozaki, M. & Hayaishi, O. (1975) Extradiol cleavage of 3-substituted catechols by an intradiol dioxygenase, pyrocatechase from a pseudomonad. *J. Biol. Chem.* **250**, 4848–4855
40. Dorn, E. & Knackmuss, H.-J. (1978) Chemical structure and biodegradability of halogenated aromatic compounds: substituent effects on 1,2-dioxygenation of catechol. *Biochem. J.* **174**, 85–94
41. Pieper, D.H., Rieneke, W., Engesser, K.-H. & Knackmuss, H.-J. (1988) Metabolism of 2,4-dichlorophenoxyacetic acid and 2-methylphenoxyacetic acid by *Alcaligenes eutrophus* JMP134. *Arch. Microbiol.* **150**, 95–102

12

Cisplatin

Elizabeth E. Trimmer[1] and John M. Essigmann[2]

Department of Chemistry and Division of Bioengineering and Environmental Health, Massachusetts Institute of Technology, Cambridge, MA 02139, U.S.A.

Introduction

The platinum co-ordination complex *cis*-diamminedichloroplatinum(II) (*cis*-DDP or cisplatin; Figure 1) was first synthesized in 1845. The useful biological effects of the compound, however, were not discovered for more than a century. In 1965, biophysicist Barnett Rosenberg was examining the effects of electrical fields on the bacterium *Escherichia coli* and observed that cells held between charged platinum electrodes grew in size but did not divide. A number of compounds had been produced during electrolysis and one, later identified as cisplatin, showed the ability to hinder cell division. The inhibitory effect on cell division suggested that cisplatin might have potential as an anti-cancer agent.

Cisplatin was approved by the United States Food and Drug Administration in 1979 and is now one of the most widely used chemotherapeutic agents for the treatment of human cancer. Cisplatin demonstrates significant activity against tumours of the ovary, bladder, lung, head and neck, but it is most strikingly effective against testicular cancer. According to recent estimates, over 90% of testicular tumours are curable, largely through the use of cisplatin-based chemotherapy. Whereas cisplatin slows the clinical course of other solid tumours, it rarely affords complete remission. In these cancers, treatment failure is often a result of drug resistance. Given the limitations of

[1]*Present address: Department of Biological Chemistry and Biophysics Research Division, University of Michigan, Ann Arbor, MI 48109-1055, U.S.A.*
[2]*To whom correspondence should be addressed.*

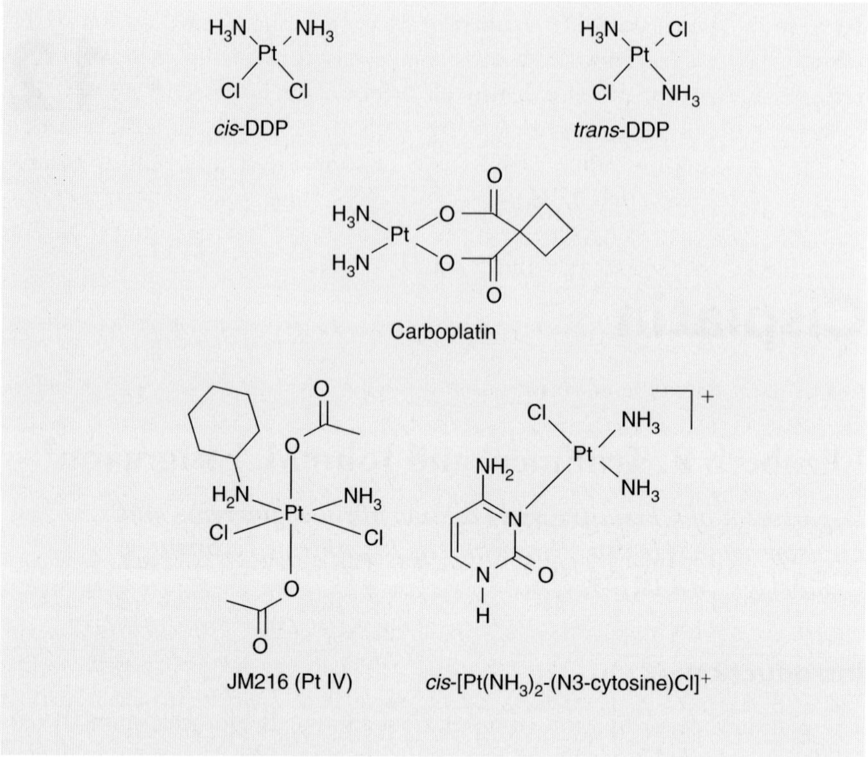

Figure 1. Structures of the anti-cancer drug cisplatin (*cis*-DDP) and other platinum compounds discussed in this Chapter

cisplatin therapy, achieving a complete understanding of the mechanism of action of cisplatin may lead to the design of novel compounds that are effective against a wider range of tumour types, to agents that ablate drug-resistant tumours, as well as to agents with fewer toxic side effects. The side effects associated with cisplatin treatment include gastrointestinal distress, kidney damage, nerve damage, hearing loss and bone-marrow suppression; some of these side effects can be alleviated partially in the clinic.

Although cisplatin's mechanism of action is not entirely known, it is believed that cisplatin exerts its cytotoxic effects through the formation of covalent adducts in which the chloride ligands of the drug are replaced by specific DNA bases. The DNA adducts of cisplatin are proposed to mediate cytotoxicity by inhibiting DNA replication and transcription and, ultimately, by activating a pathway termed programmed cell death, or apoptosis. The geometric isomer of cisplatin, *trans*-diamminedichloroplatinum(II) (*trans*-DDP; Figure 1), also binds to DNA and inhibits DNA replication, but it is at least 20-fold less toxic to cells than cisplatin and is inactive as an anti-tumour agent. This observation suggests that the cytotoxicity of cisplatin cannot be explained simply by its ability to cause DNA damage. Rather, more complex cellular and biochemical mechanisms may underlie the observed differences in toxicity between *cis*- and *trans*-DDP. Studies comparing adduct formation, repair and toxicity of the *cis* and *trans* platinum isomers have provided important insights into the mechanisms by which cisplatin, alone, displays clinical efficacy. This Chapter will summarize work on how cells differentially process platinum adducts. The review will focus on the following five areas: DNA adducts formed by *cis*- and *trans*-DDP; effects on DNA replication and transcription; repair of platinum adducts; recognition of adducts by cellular proteins; and the mechanisms of cisplatin resistance.

DNA adducts formed by *cis*- and *trans*-DDP

Cisplatin is administered intravenously to patients. In the bloodstream where the concentration of chloride ions is high (≈100 mM), cisplatin, a neutral molecule, is relatively unreactive (Figure 2). Inside the cell, however, the low ambient chloride concentration (≈4 mM) facilitates hydrolysis of the chloride ligands of the drug. Water molecules displace the two chlorides in a stepwise manner to form an aquated positively charged complex. Since water is a good leaving group, the aquated electrophilic species can react readily by ligand substitution with nucleophiles in the cell, including DNA, RNA, proteins and cellular thiols, such as glutathione and metallothioneins. The critical cellular target for cisplatin is widely believed to be DNA. The most compelling evidence in support of this view is the high sensitivity of cells with defects in DNA repair to the lethal effects of this drug. Cisplatin binds primarily at the

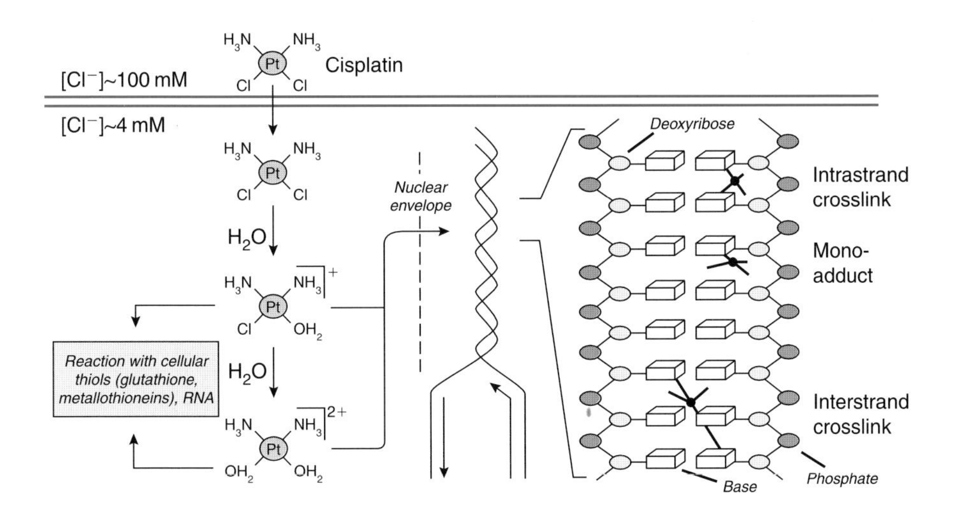

Figure 2. Schematic representation of cisplatin entering the cell and its interaction with cellular nucleophiles, including DNA
The major DNA adducts formed by cisplatin are monofunctional adducts, intrastrand crosslinks and interstrand crosslinks.

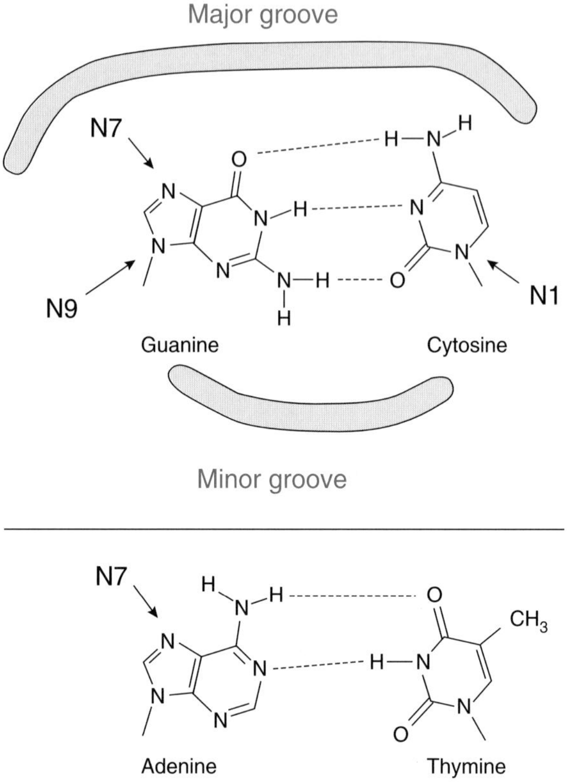

Figure 3. Watson–Crick base pairs, showing the N-7 atoms of purines, which represent the principal sites of platinum co-ordination
Also shown are the relative positions of the major and minor grooves of DNA. The N-9 and N-1 atoms of the bases are the positions at which purines and pyrimidines, respectively, are connected to the DNA deoxyribose-phosphate backbone.

N-7 positions of purine (guanine and adenine) bases, which are exposed in the major groove of the DNA double helix and are not involved in base-pair hydrogen-bonding interactions (Figure 3). Cisplatin binds to DNA in two successive steps. First, monofunctional adducts are formed by co-ordination to a single guanine or adenine. Subsequently, the remaining electrophilic centre on these monofunctional adducts will either react with a nearby purine on the same strand of the DNA to form a bifunctional intrastrand crosslink, or it will react with a purine on the complementary DNA strand to form an interstrand crosslink (Figure 2). The major DNA adducts formed by reaction of cisplatin with DNA *in vitro* include 1,2-intrastrand d(GpG) adducts between adjacent guanines (65% of detected adducts), 1,2-intrastrand d(ApG) adducts between an adjacent adenine and guanine (25%) and 1,3-intrastrand d(GpNpG) adducts between guanines separated by an intervening nucleotide (6%). Interstrand crosslinks, which form between guanines at d(G*pC)/d(G*pC)

Table 1. Comparison of DNA adducts formed by *cis*-DDP and *trans*-DDP

	cis-DDP	*trans*-DDP
Monofunctional adducts		
dG	Yes	Yes
Intrastrand crosslinks		
1,2-d(GpG)	65%	No
1,2-d(ApG)	25%	No
1,3-d(GpNpG)	6%	≈40%
Interstrand crosslinks		
d(G*pC)/d(G*pC)	2%	?
d(G*pC)/d(GpC*)	?	20%

Asterisks denote sites of platinum co-ordination.

sites (where the asterisks denote sites of platinum co-ordination), and monofunctional adducts occur at a lower frequency (1–2%; Table 1). Interestingly, 1,2-intrastrand crosslinks at d(GpA) sites are very rare. A similar DNA-adduct profile is observed in DNA isolated from the white blood cells of cancer patients following cisplatin treatment, so the pattern observed *in vitro* is also seen *in vivo*. Unfortunately, the levels of cisplatin adducts are similar in tumours and other tissues of cancer patients following cisplatin therapy, indicating that cisplatin does not localize specifically in tumour tissue.

trans-DDP (Figure 1), the clinically inactive isomer of cisplatin, binds primarily, but not exclusively, at the N-7 positions of purine bases in DNA and forms intrastrand and interstrand crosslinks, as well as monofunctional adducts. In general, the DNA-adduct spectrum of *trans*-DDP is less well characterized than that of cisplatin. The *trans*-DDP–DNA adducts include 1,3- and 1,4-intrastrand crosslinks between purine bases separated by one or two intervening nucleotides (≈40%) and interstrand d(G*pC)/d(GpC*) crosslinks between complementary guanine and cytosine bases (20%; Table 1). However, due to steric constraints, *trans*-DDP cannot form the 1,2-intrastrand adducts, and this observation has led to the suggestion that intrastrand crosslinks, which comprise over 90% of the adduct spectrum of cisplatin, are responsible for the anti-tumour activity singularly seen with cisplatin.

The structures of platinum–DNA adducts have been investigated by several methods. Gel electrophoretic-mobility studies of oligonucleotides containing site-specific platinum adducts reveal that the 1,2-d(GpG), 1,2-d(ApG) and 1,3-d(GpTpG) intrastrand crosslinks of cisplatin bend the DNA helix 34° in the direction of the major groove. The degree of unwinding induced by the 1,2-intrastrand crosslinks is 13°, whereas the 1,3-intrastrand adduct unwinds the helix to a greater extent (23°; summarized in Table 2). Recently, the structure of the 1,2-d(GpG) cisplatin intrastrand crosslink within a 12-base-pair DNA duplex was determined by X-ray crystallography [1] (Figure 4). The

Table 2. Structural alterations induced by platinum–DNA adducts

	Adducts	Degree of bending	Degree of unwinding	HMG binding
cis-DDP	Monofunctional	None	6°	No
	1,3-d(GpNpG)	34°	23°	No
	1,2-d(GpG)	34°	13°	Yes
	1,2-d(ApG)	34°	13°	Yes
	d(G*pC)/d(G*pC)	20–40°	80°	Yes
trans-DDP	1,3-d(GpNpG)	Flexible	9°	No
	d(G*pC)/d(GpC*)	Flexible, 26°	12°	No

Asterisks denote sites of platinum co-ordination.

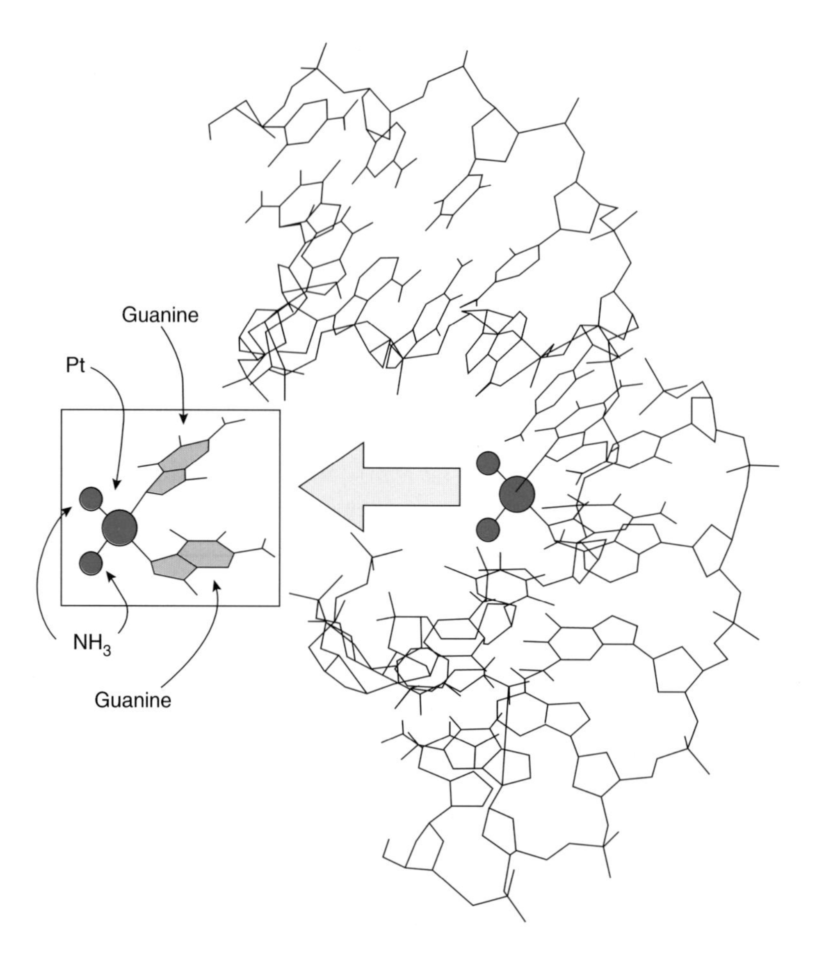

Figure 4. The structure of the 1,2-d(GpG) cisplatin intrastrand crosslink in a 12-base-pair DNA duplex

Co-ordination of platinum is to the N-7 atoms of the guanine bases. Co-ordinate data taken from [1].

platinum atom lies in the major groove of the DNA, and the DNA duplex is significantly bent towards the major groove at the site of platination. Co-ordination of the platinum to its guanine ligands destacks the bases, while base-pair hydrogen bonding is maintained. By contrast, the structure of a cisplatin interstrand d(G*pC)/d(G*pC) crosslink in duplex DNA reveals that the platinum atom lies in the minor groove of the DNA. The DNA helix is bent ≈20–40° towards the minor groove and is significantly unwound (≈80°) [2].

Structural distortions induced by the DNA adducts of *trans*-DDP are distinct from those of the cisplatin adducts (Table 2). Gel electrophoresis indicates that the 1,3-d(GpTpG) intrastrand crosslink of *trans*-DDP introduces a point of flexibility into the DNA helix rather than a directed bend, and produces a smaller degree of helix unwinding (9°) than the cisplatin-intrastrand adducts. The *trans*-DDP d(G*pC)/d(GpC*) interstrand crosslink produces a bend of 26° towards the major groove and unwinds the DNA helix by 12°. In addition, the adduct introduces a degree of flexibility to the helix [3]. Finally, monofunctional platinum adducts do not bend the DNA, but unwind the DNA helix by 6° [4].

Effects on DNA replication and transcription

It has been proposed that, like many anti-tumour agents, cisplatin exerts its cytotoxic effects through the inhibition of DNA replication. By inhibiting the synthesis of DNA in cells, cisplatin–DNA adducts would slow cell division, which in turn could provide a trigger for cell death. In this way, cisplatin would have a selective effect on rapidly dividing cells, such as tumour cells. The ability of DNA polymerases to perform synthesis on platinated DNA templates *in vitro* has been investigated in several studies. Cisplatin–DNA adducts block the progression of bacterial as well as eukaryotic DNA polymerases *in vitro*. However, DNA adducts of the inactive cisplatin isomer, *trans*-DDP are equally effective as cisplatin adducts at inhibiting the DNA polymerases. In other work, the effects of *cis*- and *trans*-DDP on replication *in vivo* was investigated using African green monkey CV-1 cells infected with the DNA tumour virus, simian virus 40 (SV40). SV40 is a model minichromosome that is replicated by the host-cell enzyme machinery and then can be recovered. SV40-infected CV-1 cells were treated with various amounts of *cis*- or *trans*-DDP and then SV40 DNA replication was monitored. It was found that the two platinum compounds were equally effective at blocking DNA replication when equal numbers of platinum adducts were bound to the SV40 DNA. These results indicate that the anti-tumour activity of cisplatin cannot be based solely on its ability to inhibit DNA replication and that additional mechanisms must be invoked in order to explain the different toxicities of *cis*- and *trans*-DDP.

Inhibition of transcription is another way that cisplatin may exert its anti-tumour effects. RNA synthesis, like DNA replication, would be more critical for a rapidly dividing tumour cell than for a stationary cell. Recent evidence suggests that, whereas the DNA adducts of *cis*- and *trans*-DDP block DNA replication equally, the adducts of these platinum isomers may inhibit RNA synthesis differentially. *In vitro* studies demonstrate that, when present on the transcribed strand, the 1,2- and 1,3-intrastrand crosslinks of cisplatin completely block the processivity of *E. coli* and wheat germ RNA polymerases. By contrast, monofunctional platinum adducts and the 1,3-d(GpTpG) intrastrand adduct of *trans*-DDP are bypassed by the polymerases [5,6]. In recent experiments *in vivo*, plasmids containing a reporter gene were modified with *cis*- or *trans*-DDP, introduced into mammalian cells, and the levels of reporter-gene transcription measured [7]. A 2–3-fold higher level of transcription was observed from plasmids modified with *trans*-DDP as compared with plasmids modified with *cis*-DDP. In particular, *cis*- and *trans*-DDP–DNA adducts are bypassed by RNA polymerase with relative efficiencies of 0–16% and 60–76%, respectively. These results suggest that the anti-tumour activity of cisplatin may be derived, at least in part, by efficient inhibition of transcription. Moreover, by inducing changes in the delicate balance of gene transcription within tumour cells, cisplatin adducts may provide a signal for cell death.

Repair of platinum adducts

Repair of cisplatin-induced DNA damage is one mechanism by which cells treated with cisplatin may increase their likelihood of survival. Nucleotide-excision repair is believed to be the main process by which bulky DNA adducts, including those formed by UV light or by cisplatin, are removed from DNA. The general repair mechanism involves recognition of the damage, incision of the DNA strand on both sides of the lesion, excision of the damaged oligonucleotide and resynthesis to fill the gap. In mammalian cells, nucleotide-excision repair requires the action of at least 30 proteins. The disease xeroderma pigmentosum (XP) results from defects in nucleotide-excision repair and is characterized by extreme sensitivity to UV light. The genes required for mammalian excision repair in rodents include those defined by the seven (A–G) complementation groups of XP as well as the excision-repair cross-complementing (*ERCC*) genes isolated from UV-sensitive rodent cell lines.

The repair of individual cisplatin–DNA adducts has been investigated in *in vitro* DNA repair assay systems employing human cell extracts. These *in vitro* repair assays either monitor DNA resynthesis after removal of the platinum damage or detect excision of the damage-containing fragment directly. Early results demonstrated that human cell extracts could carry out repair synthesis on plasmids modified globally with cisplatin. A subsequent study concluded that repair of cisplatin-modified templates results from the removal of minor adducts rather than the major 1,2-intrastrand cisplatin crosslinks. When examined directly *in vitro*, the 1,2-d(GpG) intrastrand adduct located at a specific site is less efficiently repaired (3–20-fold) by human cell extracts than the 1,3-d(GpTpG) intrastrand crosslink. Incubation of either the 1,2- or 1,3-intrastrand crosslinked substrate with human cell extracts results in the release of a 26–29-nucleotide fragment that contains the adduct. Repair of the 1,3-intrastrand adduct is also more efficient in a reconstituted repair system containing purified repair factors, suggesting that differences in the rates of repair are due to structural differences in the platinum adducts rather than to unidentified factors in the cell extracts. Taken together, these results suggest that poor repair of the major 1,2-d(GpG) intrastrand adduct may contribute to the anti-tumour activity of cisplatin.

Evidence suggests that DNA adducts of the inactive *trans*-DDP isomer may be preferentially repaired in mammalian cells compared with adducts of *cis*-DDP. This observation is in line with the lower toxicity of the *trans* isomer to cells. DNA repair synthesis assays *in vitro* indicate that human cell extracts carry out repair synthesis twice as efficiently on *trans*-DDP- than on *cis*-DDP-modified plasmids. In addition, extracts prepared from excision-repair-deficient XP complementation group A cells are unable to perform repair synthesis on *cis*-DDP- and *trans*-DDP-modified plasmids, suggesting that both types of adduct are repaired by the nucleotide excision-repair pathway. In an SV40-based *in vitro* replication assay, preincubation of *trans*-DDP-modified plasmids, but not *cis*-DDP-modified plasmids, with human cell extracts restores DNA synthesis by 30%, suggesting that an activity present in the extracts repairs the *trans*-DDP-damaged template preferentially. In support of this view, a repair assay *in vitro* demonstrated that the extracts contained a specific repair activity for *trans*-DDP adducts. Studies *in vivo*, however, have yielded conflicting results. One study reported that in SV40-infected CV-1 cells, *cis*-DDP adducts accumulate continuously over a period of 48 h, whereas *trans*-DDP adducts reach a maximum at 6 h and then levels drop dramatically. By contrast, another investigation found that DNA adducts of *cis*- and *trans*-DDP are removed from DNA at similar rates. Differences in experimental conditions, however, may not make the two studies directly comparable. It remains to be determined whether preferential repair of *trans*-DDP adducts can account for the differential toxicity between the two isomers.

Recognition of platinum adducts by cellular proteins

The structural differences among the DNA adducts formed by *cis*- and *trans*-DDP suggest that the adducts of these isomers may be processed differently by cellular proteins. To test this hypothesis, a technique called the gel-mobility-shift assay was used to search for proteins in mammalian cell extracts that bind to platinated DNA. In this assay, proteins that bind non-covalently to a radioactive DNA probe are detected by their ability to retard the migration of the probe through an electrophoresis gel. By this tool, several proteins are observed in mammalian cell extracts that bind selectively to DNA modified with cisplatin, but not to DNA adducts of the *trans*-DDP isomer. Moreover, the 1,2-d(GpG) and 1,2-d(ApG) intrastrand crosslinks of cisplatin, but not the 1,3-d(GpTpG) intrastrand adducts, are recognized by these proteins. Significantly, cisplatin–DNA binding activity is also observed in more clinically relevant extracts prepared from human testicular, ovarian and cervical tumours [8].

In parallel work, DNA modified with cisplatin was used as a probe to clone a gene that encodes a cisplatin–DNA-binding protein. In this method, a human B-cell cDNA library is constructed in the expression vector, bacteriophage λgtll. After growth of the recombinant bacteriophage on plates, isopropyl β-D-thiogalactopyranoside is added to induce expression of fusion proteins consisting of β-galactosidase linked to polypeptide sequences encoded by the cloned DNA. The screening of these fusion proteins with DNA probes modified with cisplatin led to the isolation of a positive cDNA clone encoding an 81-kDa cisplatin–DNA-binding protein. Amino acid sequence analysis revealed that the protein, termed structure-specific recognition protein 1 (SSRP1), contained a 75-amino acid conserved DNA-binding motif called the high-mobility group (HMG) domain. A family of proteins having in common the HMG DNA-binding domain has since been identified. The prototype of this family is HMG1, named for its fast mobility on SDS/PAGE. HMG1 is an abundant chromosomal protein containing two tandem HMG domains that bind to DNA cruciform structures; its function in the cell, however, is unknown. The similarity between the HMG domains of HMG1 and SSRP1 suggested that HMG1 may also recognize cisplatin-modified DNA. This notion was confirmed in gel-mobility-shift assays with purified HMG1 protein. In particular, HMG1 had the same profile of cisplatin-adduct recognition as that observed with mammalian cell extracts. Further experiments demonstrated that a single HMG domain mediates binding to cisplatin-modified DNA. To date, cisplatin–DNA-binding activity has been demonstrated for several HMG-domain proteins (summarized in Table 3). It should be emphasized that besides binding to platinated DNA, many of these proteins function in important roles in the cell, such as in the regulation of transcription.

What is the structural basis for cisplatin-adduct recognition by HMG-domain proteins? Proteins with HMG domains interact with bent DNA structures such as cruciforms and four-way junctions and induce bends in linear DNA. The interaction of the HMG domain with bent DNAs is probably due to its L-shaped tertiary structure. These observations suggest that structural distortions induced in DNA by cisplatin adducts may serve as structure-specific recognition signals for HMG-domain proteins. Moreover, the proteins appear able to distinguish the different structures induced by individual *cis*- and *trans*-DDP–DNA adducts (summarized in Table 2). For example, the major 1,2-intrastrand adducts of cisplatin, which bend the DNA 34° and unwind it 13°, are recognized by HMG-domain proteins, whereas the 1,3-intrastrand adducts, which unwind the DNA to a greater degree (23°), fail to

Table 3. HMG-domain proteins shown to bind to cisplatin-modified DNA

For details of proteins, see text.

Full-length protein	Molecular mass (kDa)	Species	Number of HMG domains	Function
SSRP1	81	Human	1	Unknown, human homologue of V(D)J recombination sequence-binding protein
Ixr1	80	Yeast	2	Transcriptional receptor
HMG1	28	Rat	2	Unknown, binds to DNA cruciforms
HMG2	26.5	Calf	2	Unknown
hUBF	97/94	Human	6	Ribosomal RNA transcription factor
tsHMG	23	Mouse	2	Testis-specific HMG protein, spermatogenesis
hSRY	24	Human	1	Testis-determining factor
MtTFA	24	Human	1	Mitochondrial transcription factor
LEF-1	44	Mouse	1	Lymphoid-enhancer-binding factor

attract these proteins. It is noteworthy that the binding of HMG-domain proteins to the 1,2-intrastrand adducts increases the DNA bend angle from 34° to 70–90°, in striking agreement with the 80° angle between the two arms of the L-shaped HMG domain.

Recognition of cisplatin adducts by HMG-domain proteins may be fortuitous; that is, the adducts may mimic naturally bent DNA substrates for these proteins. Cisplatin-adduct binding by this protein family could, however, potentially play a role in the mechanism of action of the drug. In particular, the selective affinity for the major DNA adducts of cisplatin, but not for DNA adducts of the clinically ineffective *trans*-DDP isomer, indicates that these proteins may act to enhance cisplatin cytotoxicity. One proposed model (Figure 5B) suggests that the binding of HMG-domain proteins to cisplatin adducts blocks removal of the lesions by the repair machinery of the cell. Slow repair would allow the adducts to persist on the DNA, enhancing cisplatin cytotoxicity. By contrast, DNA adducts not recognized by HMG-domain proteins would be repaired more readily, improving the chances for cell survival. Supporting this hypothesis are results from *in vivo* experiments in yeast. A yeast strain with a deletion for the HMG-domain protein Ixr1 (intrastrand crosslink recognition protein) is 2–6-fold less sensitive to cisplatin than the parental strain containing the Ixr1 protein; by contrast, sensitivity to *trans*-DDP or UV light is not altered by the absence of Ixr1. Further experiments demonstrate that the effect of Ixr1 on cisplatin sensitivity is significantly decreased in a series of yeast strains deficient in nucleotide-excision repair, establishing a direct link to repair. Additional evidence for the repair-shielding model is obtained from excision-repair assays carried out *in vitro*. The addition of HMG-domain proteins to the assay specifically inhibited excision repair of the major 1,2-d(GpG) intrastrand cisplatin crosslink, but not the 1,3-d(GpTpG) intrastrand adduct. Finally, the XP group A-complementing (XPAC) protein, which is responsible for damage recognition in nucleotide-excision repair, exhibits a lower binding affinity *in vitro* for cisplatin-modified DNA compared with the HMG proteins HMG1, Ixrl and human upstream-binding factor (hUBF). This result suggests that XPAC may not easily displace HMG-domain proteins from cisplatin–DNA adducts. Taken together, these results indicate that HMG-domain proteins may potentiate cisplatin cytotoxicity by shielding lesions from excision repair.

A second, as yet equally plausible, model (Figure 5C) proposed to explain how HMG-domain proteins may effect cisplatin toxicity is based on the fact that several HMG-domain proteins have important functions as regulators of transcription. The titration (or 'hijacking') of HMG-domain proteins away from their natural binding sites by cisplatin adducts might impair the expression of genes critical to a growing tumour cell and lead to cell death. Supporting this model are results from DNA-binding experiments *in vitro* with the HMG-domain protein hUBF, an important regulator of ribosomal RNA synthesis. The interaction of hUBF with the 1,2-d(GpG) intrastrand

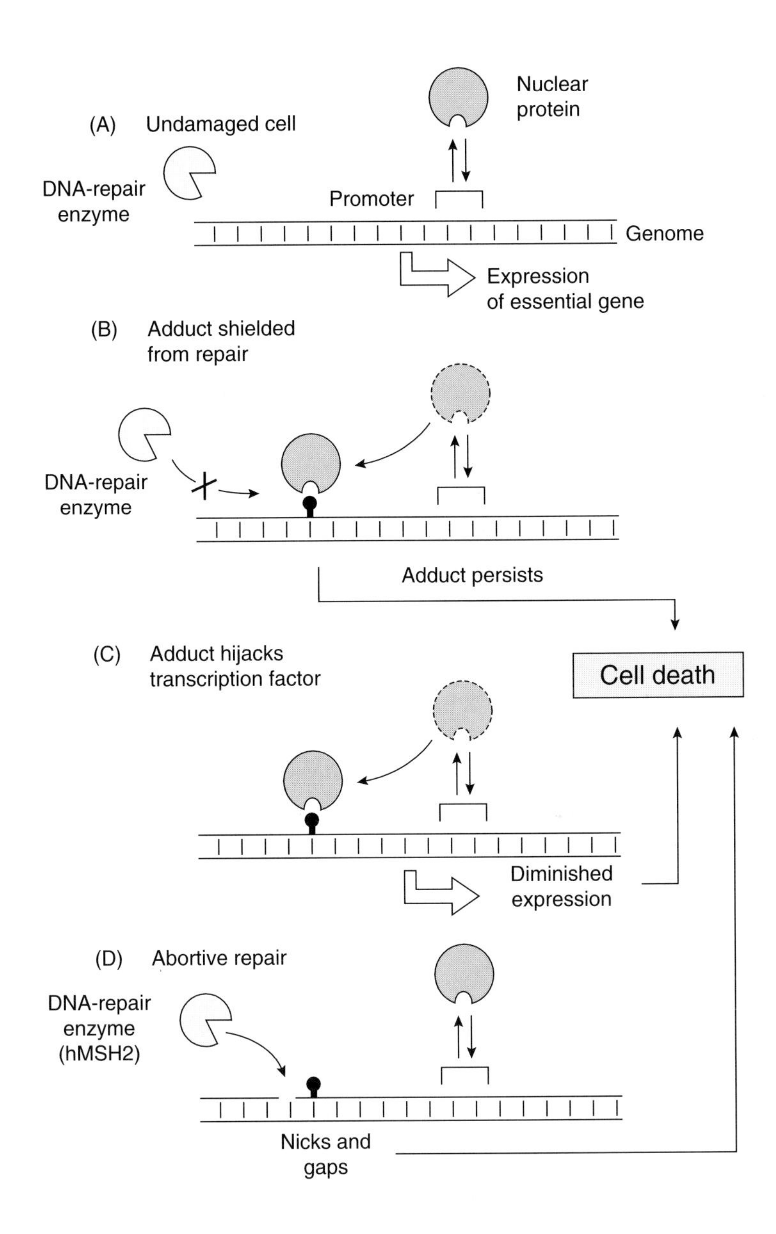

Figure 5. Models to explain how HMG domain and other nuclear proteins may enhance cisplatin cytotoxicity

(**A**) A normal cell with a nuclear protein interacting with the DNA. (**B**) Repair-shielding model. Cisplatin–DNA adducts (lollipop symbol) attract the nuclear protein and are shielded from DNA-repair enzymes, promoting the persistence of the adducts and sensitizing the cell to cisplatin. (**C**) Hijacking model. The adducts titrate the nuclear protein away from its normal site of binding, resulting in reduced expression of a critical gene. (**D**) Abortive repair model. Misdirected repair attempts at sites of cisplatin damage by mismatch DNA-repair proteins (e.g. hMSH2) may generate DNA-strand breaks that signal for cell death.

adduct of cisplatin is very favourable and rivals that of the protein for its natural promoter sequence. Moreover, the binding of hUBF to its promoter can be competed *in vitro* with a cisplatin–DNA adduct concentration (5 nM) significantly *lower* than that found in the DNA of platinum-drug-treated cancer patients (0.1–1 μM). These results suggest that cisplatin adducts may act as 'molecular decoys' in cells for HMG-domain-containing transcription factors. Thus cisplatin could have profound effects on the expression of specific genes.

As already mentioned, an important aspect of the anti-tumour activity of cisplatin is its increased effectiveness in the treatment of testicular cancer as compared with other tumour cell types. Most HMG-domain proteins are expressed in all tissue cell types and thus cannot account for the tissue-specific cytotoxicity of cisplatin. Recent evidence, however, indicates that some HMG-domain proteins are specifically expressed in the testis and thus could potentially contribute to the selective efficacy of cisplatin against testicular tumours by one or both of the mechanisms described above. In support of this view, cisplatin-adduct-binding activity has been demonstrated for a mouse testis-specific HMG-domain protein (tsHMG) [9] and the human testis-determining factor (hSRY) [10]. In the case of hSRY, the affinity of the protein for the 1,2-d(GpG) cisplatin adduct is very similar to that for its putative DNA target sequence. Furthermore, the testis-specific proteins hSRY and tsHMG inhibit excision repair of the 1,2-d(GpG) intrastrand crosslink *in vitro*.

It has also been demonstrated that the human mismatch repair protein hMSH2, which does not contain an HMG domain, binds to DNA modified with cisplatin [11,12]; importantly, this protein is overexpressed in testicular tissue [12]. As part of the mismatch DNA-repair system, hMSH2 normally functions in the recognition and repair of mismatched base pairs or small loops in DNA. The importance of mismatch repair is emphasized by the fact that mutations in mismatch repair genes are associated with over 90% of hereditary non-polyposis colorectal cancer cases. Recently, it has been proposed that mismatch repair may also be involved in the mechanism of action of cisplatin. hMSH2, in concert with other elements of the mismatch repair system, may attempt to repair cisplatin adducts and thereby generate DNA-strand breaks (Figure 5D). The latter could signal programmed cell death, or apoptosis. Accordingly, this protein, which is expressed at high levels in testis, may potentiate toxicity by this abortive repair mechanism. In addition, however, hMSH2 could participate in cell killing by binding to adducts and blocking repair, as described above for the HMG-domain proteins.

Mechanisms of cisplatin resistance

The development of drug resistance is a major factor in the failure of cisplatin-based chemotherapy to cure cancer patients. In the clinic, levels of resistance are of the order of 2–4-fold. A corresponding increase in cisplatin dose, however, would lead to severe toxicity to the patient. There are two types of cisplatin resistance, intrinsic and acquired. Intrinsic resistance is encountered with patients whose tumours are inherently resistant and do not respond to cisplatin at the time of first treatment. Acquired resistance can emerge in tumour cell populations after an initial drug response. Model systems have been valuable in the elucidation of the mechanisms of cisplatin resistance. For example, cell lines that have the ability to grow continuously in cultured media *in vitro* have been established from patient tumours. Intrinsic cisplatin resistance/sensitivity has been studied in testicular-tumour cell lines and bladder- or colon-tumour cell lines as representatives of inherently cisplatin-sensitive and -resistant tumour cell types, respectively. Cell lines that have acquired resistance by repeated or continuous exposure of the cells to cisplatin *in vitro* have been used extensively to identify mechanisms of acquired resistance. These cell lines, however, often acquire a much higher level of resistance (10–1000-fold) than tumours observed *in vivo* (2–4-fold), suggesting that these lines may not be clinically relevant models for the disease. Cell lines have also been established from tumour cells made resistant *in vivo* by cisplatin treatment, and these lines may more faithfully reflect the clinical situation. Another caveat to studying resistance in established cell lines is that some resistance mechanisms may operate only at the level of the organism, and not at the cellular level. For example, murine mammary tumours made resistant to cisplatin *in vivo* by treatment of tumour-bearing animals are surprisingly sensitive to cisplatin when the tumour cells are grown *in vitro* as continuous cultures.

Despite the caveats above, several mechanisms of cisplatin resistance have been identified by the various cellular model systems. These include: (i) reduced drug accumulation; (ii) increased drug inactivation by sulphur-containing molecules, such as glutathione and metallothionein; (ii) enhanced repair of cisplatin–DNA adducts; (iv) increased tolerance of cisplatin damage; and (v) altered expression of regulatory proteins. It is emphasized that development of resistance is a multifactorial process and, therefore, a given tumour may become resistant by one or more mechanisms.

Reduced drug accumulation

Cells can limit the formation of toxic DNA adducts by reducing the intracellular accumulation of cisplatin. Decreased drug accumulation is a consistent feature of many cisplatin-resistant cell lines, but the effects are small (2–5-fold) even when levels of resistance are high. The mechanisms underlying changes in cisplatin accumulation have not been elucidated. In fact, the pathway by which cisplatin enters cells is unclear; evidence exists both for passive diffusion and a carrier-mediated transport system. Recent studies suggest that an export pump may act to efflux cisplatin from cells. This pump is distinct, however, from the membrane drug-efflux pump encoded by the *mdr* (multidrug resistance) gene, which removes a variety of functionally different drugs from resistant cells, but is not involved in cisplatin resistance.

Drug inactivation by sulphur-containing molecules

A second mechanism by which cells may limit the number of cisplatin adducts formed is by increasing the concentration of sulphur-containing molecules that can react with the drug before it reaches the DNA (see Figure 2). Such molecules include glutathione, the most abundant thiol in the cell, and metallothioneins, small cysteine-rich proteins involved in detoxification of heavy metals. Following cisplatin treatment, a significant proportion of the total platinum content in cells is involved in a complex with either glutathione (60%) or metallothionein (25%). Elevated glutathione levels exist in cisplatin-resistant cells made resistant *in vitro* and *in vivo* and, in some cases, increased glutathione correlates with a reduction in interstrand-crosslink formation. This result supports a mechanism in which glutathione can react with monofunctional platinum adducts, preventing their closure to the more toxic bifunctional crosslinks. Glutathione is not increased, however, in other cisplatin-resistant cell lines. Similarly, metallothioneins are overexpressed in some, but not all, cell lines selected for resistance to cisplatin. Taken together, these results suggest that drug inactivation by glutathione or metallothioneins may contribute to cisplatin resistance, but additional resistance mechanisms are probably involved.

Enhanced repair of cisplatin–DNA adducts

In tumour cell lines that have acquired cisplatin resistance by incubation with the drug *in vitro*, enhanced repair of cisplatin adducts is often a major mechanism of resistance. Similarly, increased DNA repair has been observed in cell lines and primary tumour cells from patients with *in vivo*-acquired resistance. For example, an ovarian-cancer cell line established from a cisplatin-treated patient after the onset of resistance exhibits 3-fold-higher levels of DNA repair compared with a sensitive cell line derived from the same patient prior to the development of resistance.

Intrinsic resistance to cisplatin is associated with an elevated capacity for DNA repair, as evidenced in cell lines and tumours from untreated patients. Cells from non-small-cell lung tumours, which are generally resistant to cisplatin therapy, had 2–4-fold higher levels of DNA repair than cells derived from small-cell lung tumours, which are usually responsive to cisplatin treatment. Cisplatin sensitivity also correlates with DNA-repair capacity in testicular and bladder-tumour cell lines, which represent inherently cisplatin-sensitive and -resistant tumour cell types, respectively. In particular, no significant removal of the major 1,2-d(GpG) intrastrand cisplatin crosslink is detected in five testicular-tumour cell lines, whereas a bladder-tumour cell line is proficient in the repair of this adduct. These results suggest that the extreme sensitivity of testicular tumours to cisplatin may be attributed to a deficient ability to repair cisplatin–DNA adducts. Moreover, deficient DNA-repair capacity appears to be a good indicator of response to cisplatin chemotherapy.

Taken together, the results detailed above implicate enhanced repair of cisplatin–DNA adducts as a very viable mechanism of cisplatin resistance. Increases in DNA repair are usually only 2–3-fold, however, even in cells that have acquired 20–500-fold levels of resistance, suggesting that cells may activate mechanisms other than repair in order to achieve very high degrees of resistance. Enhanced repair is not always associated with cisplatin resistance. In fact, cisplatin-resistant cells have been described that are deficient in cisplatin-adduct removal [13].

Increased tolerance of cisplatin damage

An alternative strategy that cells may use to protect themselves against cisplatin is to develop an increased tolerance of cisplatin–DNA damage. Indeed, several cisplatin-resistant cell lines appear able to tolerate high levels of cisplatin–DNA adducts [14]. For resistant cells that have little or no enhanced capacity for DNA repair, this mechanism may be particularly significant. One way that resistant cells may tolerate damage is through enhanced replicative bypass of cisplatin lesions. Recently, it has been shown that calf thymus DNA polymerase β efficiently bypasses the major 1,2-d(GpG) cisplatin adduct *in vitro*. Thus, increased expression of this enzyme in cisplatin-resistant cells may enhance replicative bypass of cisplatin adducts. In support of this view, increased levels of DNA polymerase β exist in some cisplatin-resistant tumour cells. Further studies are needed in order to elucidate the mechanisms by which cisplatin-resistant cells are able to tolerate high levels of cisplatin-induced damage.

Altered expression of regulatory proteins

The altered expression of regulatory proteins by tumour cells may also contribute to cisplatin resistance. For example, cisplatin-resistant cells have demonstrated increased expression of the proto-oncogenes c-*fos*, c-*myc* and H-*ras*. In addition, the tumour-suppressor gene *p53* may play a significant role in cellular drug sensitivity. The protein encoded by *p53* is a transcriptional regulator that, in response to DNA damage, can regulate the cell cycle and, in some cell types, activate apoptosis. Significantly, functional p53 is required for the induction of apoptosis by various anti-cancer agents [15]. Thus inactivation of p53 could lead to resistance to cisplatin and other DNA-damaging agents. It is well established that cells treated by cisplatin can die by apoptosis and several studies have examined the role of p53 in cisplatin resistance/sensitivity. Lymphoma cell lines expressing mutant p53 exhibited a decreased sensitivity to cisplatin and a reduced ability to undergo apoptosis compared with lines expressing wild-type p53 [16]. Furthermore, inherently cisplatin-sensitive testicular-tumour cells, which rarely exhibit p53 mutations [17], readily undergo drug-induced apoptosis [18]. These results suggest that the extreme sensitivity of testicular tumours to cisplatin may be derived from their predominantly wild-type *p53* genotype. Taken together, the observations described above indicate that *p53* gene status is an important determinant of cisplatin resistance/sensitivity in tumour cells.

Conclusions and outlook

Testicular cancer, in contrast to most other types of solid tumour, is curable by cisplatin-based chemotherapy. Extending the success of cisplatin in the treatment of testicular cancer to other cancers will require a full understanding of the mechanism of action of the drug and its specificity for testicular tumours. Fortunately, progress has been made in recent years. DNA appears to be the critical cellular target for cisplatin, and the drug–DNA adducts, which severely distort the DNA structure, are believed to be responsible for the anti-tumour effects of the drug. It is interesting that *trans*-DDP, the geometric isomer of cisplatin, also binds to DNA, but is ineffective as a chemotherapeutic agent. This differential toxicity may stem from the formation of a different spectrum of adducts by the two compounds as well as differential processing by DNA-repair factors and other cellular proteins, such as the HMG-domain proteins. The mechanisms responsible for the development of cisplatin resistance are beginning to be elucidated. Finally, studies of testicular-tumour cells suggest that the exquisite sensitivity of this tumour cell type to cisplatin may be related to a reduced capacity to repair cisplatin-induced DNA damage. As a second factor, lethality may also be enhanced by a predominantly wild-type *p53* genotype that poises the cells to undergo programmed cell death, apoptosis, when confronted with DNA damage.

In the 20 years since the first clinical use of cisplatin, several new platinum compounds have been identified. Carboplatin (see Figure 1), a cisplatin analogue that is in use clinically, has reduced systemic toxicity compared with cisplatin, owing to its slower rate of hydrolysis. A promising new set of drugs, the platinum (IV) dicarboxylate compounds, such as JM216 (see Figure 1), has been developed recently. These compounds, which are currently undergoing clinical trials, are reduced *in vivo* to the corresponding Pt(II) species and, unlike cisplatin, can be administered orally to patients [19]. Because of their similarity in structure to cisplatin, both carboplatin and the platinum (IV) dicarboxylate drugs form a spectrum of DNA adducts that is nearly identical to that of cisplatin and, thus, they have activity against the same subset of tumours as the parent drug. At present, there is a search for platinum anti-cancer drugs that are effective against a broader range of tumour types than cisplatin. Towards this end, it has been discovered that platinum compounds of the formula *cis*-$[Pt(NH_3)_2(Am)Cl]^+$, where Am is a derivative of pyrimidine (see Figure 1), form only monofunctional DNA adducts, yet demonstrate anti-tumour activity [20]. Since these compounds form lesions that are distinct from those of cisplatin, it is possible that these new agents may have activity against tumours that are inherently resistant to cisplatin. We eagerly await the elucidation of the mechanism of action of this class of platinum compounds.

Finally, it is evident from the foregoing discussion that much has been learned about the unusual ways that cells respond to cisplatin. For example, the binding of proteins to adducts with the outcome of enhanced lethality is a novel notion that has stimulated thinking towards the design of a new generation of anti-cancer drugs. Recently, progress has been made with the synthesis of novel toxins that selectively kill cancer cells, based upon the lessons learned from cisplatin [21].

Summary

- *Cisplatin is a widely used anti-cancer drug that is exceptionally effective against testicular cancer. trans-DDP, the geometric isomer of cisplatin, is ineffective as a chemotherapeutic agent.*
- *The anti-tumour activity of cisplatin is generally attributed to its formation of DNA adducts, both intrastrand and interstrand crosslinks, which induce structural distortions in DNA.*
- *The DNA adducts of cisplatin are thought to mediate its cytotoxic effects by inhibiting DNA replication and transcription and, ultimately, by inducing programmed cell death, or apoptosis.*
- *The adducts of both cis- and trans-DDP are removed from DNA by the nucleotide-excision-repair pathway.*
- *Cellular proteins possessing certain DNA-binding motifs, including the HMG domain, bind selectively to DNA modified by cisplatin, but not to DNA adducts of trans-DDP; evidence suggests a possible role for these proteins in modulating cisplatin cytotoxicity.*
- *Both intrinsic and drug-induced resistance often limit the success of cisplatin; several specific mechanisms of cisplatin resistance have been identified.*

Our studies on cisplatin have been supported by grants CA52127 and CA34992 from the National Institutes of Health.

Further reading

Andrews, P.A. & Howell, S.B. (1990) Cellular pharmacology of cisplatin: perspectives on mechanisms of acquired resistance. *Cancer Cells* **2**, 35–43

Bruhn, S.L., Toney, J.H. & Lippard, S.J. (1990) Biological processing of DNA modified by platinum compounds. *Prog. Inorg. Chem.* **38**, 477–516

Chu, G. (1994) Cellular responses to cisplatin. *J. Biol. Chem.* **269**, 787–790

Mello, J.A., Trimmer, E.E., Kartalou, M. & Essigmann, J.M. (1998) The conflicting roles of mismatch repair and nucleotide excision repair in cellular susceptibility to anticancer drugs. *Nucleic Acids Mol. Biol.* **12**, 249–274

Sherman, S.E. & Lippard, S.J. (1987) Structural aspects of platinum anticancer drug interactions with DNA. *Chem. Rev.* **87**, 1153–1181

Whitehead, J.P. & Lippard, S.J. (1996) Proteins that bind to and mediate the biological activity of platinum anticancer drug-DNA adducts. *Met. Ions Biol. Syst.* **32**, 687–726

Wood, R.D. (1996) DNA repair in eukaryotes. *Annu. Rev. Biochem.* **65**, 135–167

References

1. Takahara, P.M., Frederick, C.A. & Lippard, S.J. (1996) Crystal structure of the anticancer drug cisplatin bound to duplex DNA. *J. Am. Chem. Soc.* **118**, 12309–12321
2. Huang, H., Zhu, L., Reid, B.R., Drobny, G.P. & Hopkins, P.B. (1995) Solution structure of a cisplatin-induced DNA interstrand cross-link. *Science* **270**, 1842–1845
3. Brabec, V., Sip, M. & Leng, M. (1993) DNA conformational change produced by the site-specific interstrand cross-link of *trans*-diamminedichloroplatinum (II). *Biochemistry* **32**, 11676–11681
4. Keck, M.V. & Lippard, S.J. (1992) Unwinding of supercoiled DNA by platinum-ethidium and related complexes. *J. Am. Chem. Soc.* **114**, 3386–3390
5. Corda, Y., Job, C., Anin, M-F., Leng, M. & Job, D. (1991) Transcription by eucaryotic and procaryotic RNA polymerases of DNA modified at a d(GG) or a d(AG) site by the antitumour drug *cis*-diamminedichloroplatinum(II). *Biochemistry* **30**, 222–230
6. Corda, Y., Job, C., Anin, M.-F., Leng, M. & Job, D. (1993) Spectrum of DNA-platinum adduct recognition by prokaryotic and eukaryotic DNA-dependent RNA polymerases. *Biochemistry* **32**, 8582–8588
7. Mello, J.A., Lippard, S.J. & Essigmann, J.M. (1995) DNA adducts of *cis*-diamminedichloroplatinum (II) and its trans isomer inhibit RNA polymerase II differentially *in vivo*. *Biochemistry* **34**, 14783–14791
8. Bissett, D., McLaughlin, K., Kelland, L.R. & Brown, R. (1993) Cisplatin-DNA damage recognition proteins in human tumour extracts. *Br. J. Cancer* **67**, 742–748
9. Ohndorf, U.M., Whitehead, J.P., Raju, N.L. & Lippard, S.J. (1997) Binding of tsHMG, a mouse testis-specific HMG-domain protein, to cisplatin-DNA adducts. *Biochemistry* **36**, 14807–14815
10. Trimmer, E.E., Zamble, D.B., Lippard, S.J. & Essigmann, J. (1998) Human testis-determining factor SRY binds to the major DNA adduct of cisplatin and a putative target sequence with comparable affinities. *Biochemistry* **37**, 352–362
11. Duckett, D.R., Drummond, J.T., Murchie, A.I.H., Reardon, J.T., Sancar, A., Lilley, D.M.J. & Modrich, P. (1996) Human MutSα recognizes damaged DNA base pairs containing O^6-methylguanine, O^4-methylthymine, or the cisplatin-d(GpG) adduct. *Proc. Natl. Acad. Sci. U.S.A.* **93**, 6443–6447
12. Mello, J.A., Acharya, S., Fishel, R. & Essigmann, J.M. (1996) The mismatch-repair protein hMSH2 binds selectively to DNA adducts of the anticancer drug cisplatin. *Chem. Biol.* **3**, 579–589
13. Shellard, S.A., Hosking, L.K. & Hill, B.T. (1991) Anomalous relationship between cisplatin sensitivity and the formation and removal of platinum-DNA adducts in two human ovarian cell lines *in vitro*. *Cancer Res.* **51**, 4557–4564
14. Johnson, S.W., Laub, P.B., Beesley, J.S., Ozols, R.F. & Hamilton, T.C. (1997) Increased platinum-DNA damage tolerance is associated with cisplatin resistance and cross-resistance to various chemotherapeutic agents in unrelated human ovarian cancer cell lines. *Cancer Res.* **57**, 850–856
15. Lowe, S.W., Ruley, H.E., Jacks, T. & Housman, D.E. (1993) p53-Dependent apoptosis modulates the cytotoxicity of anticancer agents. *Cell* **74**, 957–967
16. Fan, S., El-Deiry, S., Bae, I., Freeman, J., Jondle, D., Bhatia, K., Fornace, Jr., A.J., Magrath, I., Kohn, K.W. & O'Connor, P.M. (1994) *p53* Gene mutations are associated with decreased sensitivity of human lymphoma cells to DNA damaging agents. *Cancer Res.* **54**, 5824–5830
17. Peng, H.-Q., Hogg, D., Malkin, D., Bailey, D., Gallie, B.L., Bulbul, M., Jewett, M., Buchanan, J. & Gross, P.E. (1993) Mutations of the p53 gene do not occur in testis cancer. *Cancer Res.* **53**, 3574–3578

18. Chresta, C.M., Masters, J.R.W. & Hinkman, J.A. (1996) Hypersensitivity of human testicular tumours to etoposide-induced apoptosis is associated with functional p53 and a high Bax:Bcl-2 ratio. *Cancer Res.* **56** 1834–1841
19. Kelland, L.R., Murrer, B.A., Able, G., Giandomenico, C.M., Mistry, P. & Harrap, K.R. (1992) Ammine/amine platinum (IV) dicarboxylates: a novel class of platinum complex exhibiting selective cytotoxicity to intrinsically cisplatin-resistant human ovarian carcinoma cell lines. *Cancer Res.* **52**, 822–828
20. Hollis, L.S., Amundsen, A.R. & Stern, E.W. (1989) Chemical and biological properties of a new series of cis-diammineplatinum(II) antitumour agents containing three nitrogen donors: cis-[Pt(NH3)2(N-donor)Cl]+. *J. Med. Chem.* **32**, 128–136
21. Rink, S.M., Yarema, K.J., Paige, L.A., Tadayoni, M., Solomon, M., Essigmann, J.M. & Croy, R.G. (1996) Synthesis and biological activity of DNA damaging agents that form decoy binding sites for the estrogen receptor. *Proc. Natl. Acad. Sci. U.S.A.* **93**, 15063–15068

Subject index

A

acetogenesis, 150–151
adenosylcobalamin, 143–147, 151
adenosyl radical, 143
amine oxidase, 160, 165
amino mutase, 145–146
anti-cancer agent, 191, 207
antimony resistance, 9–12
apoptosis, 204
aromatic compound, 32, 33, 177
aromatic-ring hydroxylase, 41, 43
aromatic-ring oxygenation, 35
arsenic resistance, 9–12
ars operon, 9–12
ATPase, 3–5
ATPase inhibitor, 7
autoxidation, 95

B

B_{12}-mediated catalysis, 141
bacterial
  copper transporter, 6
  detoxification, 17–30
  molybdenum oxotransferase, 133–134
  soft-metal resistance, 1–15
benzoic acid, 174
biodegradative pathway, 32
bioinformatics, 73
bioremediation, 66, 177

C

cadmium resistance, 7–9
cad operon, 7–8
calcium-extrusion pump, 3
cancer
  anti-cancer agent, 191, 207
  hereditary non-polyposis colorectal, 204
  testicular, 191, 204
carbon monoxide binding, 92–93
carbon monoxide/oxygen affinity, 92
carbon-skeleton mutase, 146–147
carboplatin, 208
catechol dioxygenase, 34–36, 173–189
cation translocation, 3
chemotherapy, 191
chloroperoxidase, 54, 57–59, 60
chlorophyll, 140
cisplatin, 191–211
cobalamin (coenzyme B_{12}), 139–154
cobalt, 141
cofactor
  biogenesis, 168–170
  copper, 5
  pterin, 42, 126, 128
  pyranopterin, 127
  quino-cofactor, 163
  redox, 162–164
CopA/CopB, 5–6
copper, 5–6, 156, 161–162
copper oxidase, 155, 160–161, 166
copper P-type ATPase, 5
corrin ring, 140, 151
cross-reaction rate constant, 113
cytochrome *c*, 109, 112, 117, 120, 121
cytochrome *c* peroxidase, 56, 57, 120, 121
cytochrome P450, 58, 64–65, 71–83
cytotoxicity, 192, 203
czc operon, 8–9

D

deaminase, 145
dehydratase, 145
detoxification, 17–30
dicopper(II)–peroxo complex, 95
di-iron-oxo mono-oxygenase, 45
dioxygen, 81, 86, 156, 157, 159
dioxygenase, 33, 36–41, 75, 173, 182–183
DMSO reductase, 134, 135
DNA
  adduct, 193–197, 200, 205, 206
  damage, 207
  repair, 198, 206
  replication, 197
drug
  metabolism, 74
  resistance, 204–207

E

electrochemistry, 106–112
electron donor/acceptor, 117, 118
electron transfer, 27, 65, 101–124
enzyme–substrate complex, 179
EPR spectroscopy, 179, 180, 185
extended X-ray absorption fine structure spectroscopy, 179

F

Fe(II)–pterin-dependent hydroxylase, 41–43
ferric haem, 77
flavin, 107, 128, 135
flavin hydroquinone, 115
flavoprotein dehydrogenase, 75
Franck–Condon principle, 103, 105
free radical, 162

G

galactose oxidase, 160, 167
gel-mobility-shift assay, 199
glutaredoxin, 147
glutathione, 206
glyoxal oxidase, 160

H

H^+/K^+-ATPase, 3
haem, 77, 128, 135, 140
haem-containing peroxidase, 51
haemerythrin, 85, 93–95
haemocyanin, 85, 95–96
haemoglobin, 85, 90
haemoprotein, 78
halogenated hydrocarbon, 176
heavy-metal toxicity, 7
hereditary non-polyposis colorectal cancer, 204
high-mobility group domain, 200, 202, 203
homocysteine, 148
hormone biosynthesis, 74
horseradish peroxidase, 52–64
hydrogen peroxide, 159
hydroxylase, 41–43
hydroxylation, 130

I

inhibitor binding, 181
interstrand crosslink, 194
intramolecular electron transfer, 118
intrastrand crosslink, 194, 196
iron
  complex, 173, 185
  di-iron-oxo mono-oxygenase, 45
  Fe–O_2 unit, 90–91
  mononuclear centre, 38–40
  non-haem centre, 33–34, 38, 128
  non-haem oxygenase, 35
isomerase, 143–147, 151

L

Lewis acidity, 186
lysine tyrosyl quinone, 163
lysyl oxidase, 160

M

mercuric ion reductase (MerA), 25, 26
mercury, 17–28
mer operon, 17–22
metabolic transformation, 74
metal
  centre, 107, 113, 115
  cluster, 126
  resistance, 1–15, 17–30
metal–dioxygen adduct, 86
metalloprotein, 113, 116–117
metallothionein, 206
methane mono-oxygenase, 44, 45
methanogenesis, 150–151
methionine synthase, 142, 148–150
methylcobalamin, 148–151
methylmalonyl-CoA mutase, 142
methyl transferase, 150–151
midpoint potential, 108, 110
molecular orbital, 86
molybdenum, 125–137
molybdenum hydroxylase, 127–130
molybdenum oxotransferase, 131–134
mononuclear Fe(II) dioxygenase, 36–41
mononuclear iron centre, 38–40
mono-oxygenase, 33, 40, 44, 45, 64, 75
Mössbauer spectroscopy, 180
myoglobin, 85, 90, 117
myohaemerythrin, 85, 93–95

N

nitrate reductase, 131
nitrogenase, 126
non-haem iron, 33–34, 38, 128
non-haem iron oxygenase, 35
nucleophilic attack, 148
nucleotide, 194
nucleotide-excision repair, 198

O

O–O bond cleavage, 80
O–O distance, 86
O–O stretching, 86, 96
ordination environment, 177
organomercurial resistance, 24
oxidase, 159
oxidation–reduction, 107, 108, 116, 119, 155
oxotransferase, 131–134
oxygen
  acceptor/donor, 38, 61, 132
  activation, 33–34, 97, 155, 175, 176, 179
  binding, 92–93, 94, 96
  -carrying protein, 85, 87–90
  insertion, 77
  metabolism, 75
  molecular dioxygen, 81, 86, 156, 157, 159
  reactive oxygen species, 5
  redox, 156–159

P

peroxidase, 52–65
peroxide, 176
phenoxyl radical, 164
phthalate dioxygenase system, 37
phylogeny
  of cytochrome P450, 73–77
  of P-type ATPase, 3–5
platinum, 193, 199
porphyrin, 140
potential energy, 104
protein–protein electron transfer, 119–121
proteomics, 73
protoporphyrin IX, 90
pterin cofactor, 42, 126, 128
P-type ATPase, 3–5
pyranopterin cofactor, 127

Q

quino-cofactor, 163

R

rate constant
  cross-reaction, 113
  electron transfer, 103, 104
  self-exchange, 113
reactive oxygen species, 5
redox
  chemistry, 156–159, 176
  cofactor, 162–164
  transfer, 76
redox-active protein, 44, 127, 131, 164
reduction potential, 27
Reiske oxygenase, 36–41
repair shielding, 202
resistance
  drug, 204–207
  organomercurial, 24
  metal, 1–15, 17–30
Resonance Raman spectroscopy, 177
ribonucleotide reductase, 147
16S ribosomal RNA, 73

S

self-exchange rate constant, 113
simian virus 40, 197
sodium pump, 3
soft-metal P-type ATPase, 3–5
soft-metal resistance, 1–15
spectroelectrochemistry, 107, 110
spectroscopy
  EPR, 179, 180, 185
  extended X-ray absorption fine structure, 178
  Mössbauer, 180
  Resonance Raman, 177
  X-ray, 181
spin barrier, 157
substrate activation/binding, 181
sulphite oxidase, 131, 132, 133
sulphur, 206
superoxide, 159, 176

T

testicular cancer, 191, 204
thioredoxin, 147
thiyl radical, 170
toluene mono-oxygenase, 44
topaquinone, 163, 169
toxicity of metal (*see* metal resistance)
transcription, 20, 197
trans-diamminedichloroplatinum(II), 192, 195, 199
tumour cell, 207
tyrosine–cysteine, 163, 166, 168

U

upstream-binding factor, 202

V

vitamin B_{12}, 139

X

xanthine dehydrogenase, 127
xanthine oxidase, 127, 129, 130
xeroderma pigmentosum, 198
X-ray
  crystallography, 185, 195
  spectroscopy, 181

Z

zinc
  ion, 148
  resistance, 7–9
ZntA, 7–8